Statistical Arbitrage

Founded in 1807, John Wiley & Sons is the oldest independent publishing company in the United States. With offices in North America, Europe, Australia, and Asia. Wiley is globally committed to developing and marketing print and electronic products and services for our customers' professional and personal knowledge and understanding.

The Wiley Finance series contains books written specifically for finance and investment professionals as well as sophisticated individual investors and their financial advisors. Book topics range from portfolio management to e-commerce, risk management, financial engineering, valuation, and financial instrument analysis, as well as much more.

For a list of available titles, visit our Web site at www.WileyFinance.com.

Statistical Arbitrage

Algorithmic Trading Insights and Techniques

ANDREW POLE

John Wiley & Sons, Inc.

Published by John Wiley & Sons, Inc., Hoboken, New Jersey.
Published simultaneously in Canada.

Wiley Bicentennial logo: Richard J. Pacifico.

Library of Congress Cataloging-in-Publication Data:

Pole, Andrew, 1961–
Statistical arbitrage : algorithmic trading insights and techniques / Andrew Pole.
p. cm. — (Wiley finance series)
Includes bibliographical references and index.
ISBN 978-0-470-13844-1 (cloth)
1. Pairs trading. 2. Arbitrage---Mathematical models. 3. Speculation--Mathematical models. I. Title.
HG4661.P65 2007
332.64'5 — dc22

2007026257

ISBN 978-0-470-13844-1

10 9 8 7 6 5 4 3 2 1

To Eliza and Marina

Contents

Preface xiii

Foreword xix

Acknowledgments xxiii

CHAPTER 1
Monte Carlo or Bust 1

Beginning 1
Whither? And Allusions 4

CHAPTER 2
Statistical Arbitrage 9

Introduction 9
Noise Models 10
Reverse Bets 11
Multiple Bets 11
Rule Calibration 12
Spread Margins for Trade Rules 16
Popcorn Process 18
Identifying Pairs 20
Refining Pair Selection 21
Event Analysis 22
Correlation Search in the Twenty-First Century 26
Portfolio Configuration and Risk Control 26
Exposure to Market Factors 29
Market Impact 30
Risk Control Using Event Correlations 31
Dynamics and Calibration 32
Evolutionary Operation: Single Parameter Illustration 34

CHAPTER 3
Structural Models **37**

Introduction 37
Formal Forecast Functions 39
Exponentially Weighted Moving Average 40
Classical Time Series Models 47
Autoregression and Cointegration 47
Dynamic Linear Model 49
Volatility Modeling 50
Pattern Finding Techniques 51
Fractal Analysis 52
Which Return? 52
A Factor Model 53
Factor Analysis 54
Defactored Returns 55
Prediction Model 57
Stochastic Resonance 58
Practical Matters 59
Doubling: A Deeper Perspective 61
Factor Analysis Primer 63
Prediction Model for Defactored Returns 65

CHAPTER 4
Law of Reversion **67**

Introduction 67
Model and Result 68
The 75 percent Rule 68
Proof of the 75 percent Rule 69
Analytic Proof of the 75 percent Rule 71
Discrete Counter 73
Generalizations 73
Inhomogeneous Variances 74
Volatility Bursts 75
Numerical Illustration 76
First-Order Serial Correlation 77
Analytic Proof 79
Examples 82
Nonconstant Distributions 82
Applicability of the Result 84
Application to U.S. Bond Futures 85

Summary 87
Appendix 4.1: Looking Several Days Ahead 87

CHAPTER 5
Gauss Is Not the God of Reversion **91**

Introduction 91
Camels and Dromedaries 92
Dry River Flow 95
Some Bells Clang 98

CHAPTER 6
Interstock Volatility **99**

Introduction 99
Theoretical Explanation 103
Theory versus Practice 105
Finish the Theory 105
Finish the Examples 106
Primer on Measuring Spread Volatility 108

CHAPTER 7
Quantifying Reversion Opportunities **113**

Introduction 113
Reversion in a Stationary Random Process 114
Frequency of Reversionary Moves 117
Amount of Reversion 118
Movements from Quantiles Other Than the Median 135
Nonstationary Processes: Inhomogeneous Variance 136
Sequentially Structured Variances 136
Sequentially Unstructured Variances 137
Serial Correlation 138
Appendix 7.1: Details of the Lognormal Case in Example 6 139

CHAPTER 8
Nobel Difficulties **141**

Introduction 141
Event Risk 142
Will Narrowing Spreads Guarantee Profits? 144
Rise of a New Risk Factor 145

Redemption Tension 148
Supercharged Destruction 150
The Story of Regulation Fair Disclosure (FD) 150
Correlation During Loss Episodes 151

CHAPTER 9
Trinity Troubles 155

Introduction 155
Decimalization 156
European Experience 157
Advocating the Devil 158
Stat. Arb. Arbed Away 159
Competition 160
Institutional Investors 163
Volatility Is the Key 163
Interest Rates and Volatility 165
Temporal Considerations 166
Truth in Fiction 174
A Litany of Bad Behavior 174
A Perspective on 2003 178
Realities of Structural Change 179
Recap 180

CHAPTER 10
Arise Black Boxes 183

Introduction 183
Modeling Expected Transaction Volume and Market Impact 185
Dynamic Updating 188
More Black Boxes 189
Market Deflation 189

CHAPTER 11
Statistical Arbitrage Rising 191

Catastrophe Process 194
Catastrophic Forecasts 198
Trend Change Identification 200
Using the Cuscore to Identify a Catastrophe 202
Is It Over? 204
Catastrophe Theoretic Interpretation 205
Implications for Risk Management 209

Sign Off 211
Appendix 11.1: Understanding the Cuscore 211

Bibliography **223**

Index **225**

Preface

These pages tell the story of statistical arbitrage. It is both a history, describing the first days of the strategy's genesis at Morgan Stanley in the 1980s through the performance challenging years of the early twenty-first century, and an exegesis of how and why it works. The presentation is from first principles and largely remains at the level of a basic analytical framework. Nearly all temptation to compose a technical treatise has been resisted with the goal of contributing a work that will be readily accessible to the larger portion of interested readership. I say "nearly all": Chapter 7 and the appendix to Chapter 11 probably belong to the category of "temptation not resisted." Much of what is done by more sophisticated practitioners is discussed in conceptual terms, with demonstrations restricted to models that will be familiar to most readers. The notion of a pair trade—the progenitor of statistical arbitrage—is employed to this didactic end rather more broadly than actual trading utility admits. In adopting this approach, one runs the risk of the work being dismissed as a pairs trading manual; one's experience, intent, and aspirations for the text are more extensive, but the inevitability of the former is anticipated. In practical trading terms, the simple, unelaborated pair scheme is no longer very profitable, nonetheless it remains a valuable tool for explication, retaining the capacity to demonstrate insight, modeling, and analysis while not clouding matters through complexity. After a quarter century in the marketplace, for profitable schemes beyond paper understanding and illustration, one needs to add some structural complexity and analytical subtlety.

One elaboration alluded to in the text is the assembling of a set of similar pairs (without getting into much detail on what metrics are used to gauge the degree of similarity), often designated as a group. Modeling such groups can be done in several ways, with some practitioners preferring to anchor a group on a notional archetype, structuring forecasts in terms of deviation of tradable pairs from the archetype; others create a formal implementation of the cohort as a gestalt or a synthetic instrument. Both of those approaches, and others, can be formally analyzed as a hierarchical model, greatly in vogue (and greatly productive of insight and application) in mainstream statistical thinking for two decades; add to the standard static structure the dynamic element in a time series setting and one is very quickly building an analytical structure of greater sophistication than routinely used as the didactic tool in this book. Nonetheless, all such modeling developments rely on the insight and techniques detailed herein.

Those readers with deeper knowledge of mathematical and statistical science will, hopefully, quickly see where the presentation can be taken.

Maintaining focus on the structurally simple pair scheme invites readers to treat this book as an explicit "how to" manual. From this perspective, one may learn a reasonable history of the *what* and the *how* and a decent knowledge of *why it is possible*. Contemporary successful execution will require from the reader some additional thought and directed exploration as foregoing remarks have indicated. For that task, the book serves as a map showing major features and indicating where the reader must get out a compass and notebook. The old cartographers' device "Here be dragons" might be usefully remembered when you venture thus.

The text has, unashamedly, a statistician's viewpoint: Models can be useful. Maintaining a model's utility is one theme of the book. The statistician's preoccupation with understanding variation—the appreciation of the knowledge that one's models are wrong, though useful, and that the nature of the wrongness is illuminated by the structure of "errors" (discrepancies between observations and what a model predicts) is another theme of the book. Or, rather, not a distinct theme, but an overriding, guiding context for the material.

The notion of a pair trade is introduced in Chapter 1 and elaborated upon in Chapter 2. Following explication and exemplification, two simple theoretical models for the underlying phenomenon exploited by pairs, reversion, are proposed. These models are used throughout the text to study what is possible, illuminate how the possibilities might be exploited, consider what kinds of change would have negative impact on exploitation, and characterize the nature of the impact. Approaches for selecting a universe of instruments for modeling and trading are described. Consideration of change is introduced from this first toe dipping into analysis, because temporal dynamics underpin the entirety of the project. Without the dynamic there is no arbitrage.

In Chapter 3 we increase the depth and breadth of the analysis, expanding the modeling scope from simple observational rules[1] for pairs to formal statistical models for more general portfolios. Several popular models for time series are described but detailed focus is on weighted moving averages at one extreme of complexity and factor analysis at another, these extremes serving to carry the message as clearly as we can make it. Pair spreads are referred to throughout the text serving, as already noted, as the simplest practical illustrator of the notions discussed. Where necessary to make our urgencies sensible, direct mention is made of other aspects of the arbitrageur's concern, including portfolio optimization and factor exposures. For the most part though, incursions into multivariate territory are avoided. Volatility modeling (and the fascinating idea of stochastic resonance) are treated separately here and in Chapter 6; elsewhere discussion is subsumed in that of the mean forecast process.

Chapter 4 presents a probability theorem that illuminates the prevalence of price moves amenable to exploitation by the simple rules first applied in the late 1980s. The insight of this result guides evaluation of exploitation strategies. Are results borne of brilliance on the part of a modeler or would a high school graduate perform similarly because the result is driven by structural dynamics, long in the public domain, revealed by careful observation alone? Many a claim of a "high" proportion of winning bets by a statistical arbitrageur has more to do with the latter than any sophistication of basic spread modeling or (importantly) risk management. When markets are disrupted and the conditions underlying the theoretical result are grossly violated, comparative practitioner performance reveals much about basic understanding of the nature of the process

[1]There is no pejorative intent in the use of the term: The rules were effective. Statistical content was limited to measurement of range of variation; no distributional study, model formulation, estimation, error analysis, or forecasting was undertaken prior to milking the observational insight. Those activities came soon enough—after the profits were piling up. With the expanded statistical study, adding trading experience to historical data, came insight into subtleties of the stock price motions exploited and the market forces driving repetitious occurrence of opportunities.

being exploited. Knowledge of the theoretical results often reveals itself more when assumptions are violated than when things are hunky dory and managers with solid understanding and those operating intellectually blind generate positive returns in equal measure. (Tony O'Hagan suggested that the basic probability result is long known, but I have been unable to trace it. Perhaps the result is too trivial to be a named result and exists as a simple consequence, a textbook exercise, of basic distribution theory. No matter, the implication remains profoundly significant to the statistical arbitrage story.)

Chapter 5 critiques a published article (whose authors remain anonymous here to protect their embarrassment) to clarify the broad conditions under which the phenomenon of reversion occurs. A central role for the normal distribution is dismissed. The twin erroneous claims that (a) a price series must exhibit a normal marginal distribution for reversion to occur, and (b) a series exhibiting a normal marginal distribution necessarily exhibits reversion are unceremoniously dispelled. There is reversion anywhere and everywhere, as Chapter 4 demonstrates.

Chapter 6 answers the question, important for quantifying the magnitude of exploitable opportunities in reversion gambits, "How much volatility is there in a spread?"

Chapter 7 is for the enthusiast not easily dissuaded by the presence of the many hieroglyphs of the probability calculus. Anyone with a good first course in probability theory can follow the arguments, and most can manage the detailed derivations, too. The mechanics are not enormously complicated. Some of the conceptual distinctions may be challenging at first—read it twice! The effort will be repaid as there is significant practical insight in the examples considered at length. Knowledge of how close theoretical abstractions come to reflecting measurable features of actual price series is invaluable in assessing modeling possibilities and simulation or trading results. Notwithstanding that remark, it is true that the remainder of the book does not rely on familiarity with the material in Chapter 7. While you may miss some of the subtlety in the subsequent discussions, you will not lack understanding for omitting attention to this chapter.

Chapters 8 through 10 might have been labeled "The Fall," as they characterize the problems that beset statistical arbitrage beginning in 2000 and directly caused the catastrophic drop in return during 2002–2004. An important lesson from this history is that there was not a single condition or set of conditions that abruptly changed in 2000 and thereby eliminated forecast performance of statistical arbitrage models. What a story that would be! Far more dramatic than the prosaic reality, which is a complex mix of multiple causes and timings. All the familiar one liners, including decimalization, competition, and low volatility, had (and have) their moment, but none individually, nor the combination, can have delivered a blow to financial markets. Fundamentally altering the price dynamics of markets in ways that drastically diminish the economic potential in reversion schemes, mining value across the spectrum from the very high frequency hare of intra-day to the venerable tortoise of a month or more, requires a more profound explanation.

Change upon change upon change cataloged in Chapter 9 is at the root of the dearth of return to statistical arbitrage in 2002–2004. (Performance deterioration in 2000–2002 was evident but limited to a subset of practitioners.) This unusual episode in recent U.S. macroeconomic history is over, but the effects linger in the financial markets reflecting emergent properties of the collective behavior of millions of investors; and surely those investors continue to embody, no matter how lingering, those changes and the causes thereof.

The shift of trading from the floor of the New York Stock Exchange to internal exchanges, in the guise of computer algorithms designed by large brokerage houses and investment banks, has cumulatively become a change with glacier-like implacability. Slow. Massive. Irresistible. Crushing. Reforming.[2] A frequently remarked facet of the evolving dynamics is the decline of market volatility. Where has market volatility gone? In large part the algorithms have eaten it. Reduce the voice of a single participant yelling in a crowd and the babel is unaffected. Quite a significant proportion of participants and the reduced babel is oddly deafening. Now that computer programs (Chapter 10) "manage" over 60 percent of U.S. equity trades among "themselves" the extraodinary result is akin to administering a dose of ritalin to the hyperactive market child. In the commentary on low volatility two themes stand out: one is a lament over the lack

[2]One major structural consequence, fed also by technical advance in the credit markets and the development of Exchange Traded Funds, is literally the forming anew of patterns of price behavior detemined by the interaction of computer algorithms as agents for share dealings. In addition to this re-forming, reform is simultaneously underway with changes to Securities Exchange Commission regulations and NYSE rules.

of Keynes' animal spirits, a concern that the entrepreneurial genius of America is subdued even as Asian giants are stirring; the other is a fear that investors have forgotten the risks inherent in investment decisions, that inadvisable decisions are therefore being made that will have negative consequences in the near future. The inconsistency in those two characterizations is stark, but it can be rationalized. Contrary to the first notion, the spirit is quite animated—with a billion and a half shares changing ownership daily on the NYSE mart alone, what other conclusion should one draw? There is plenty of spirit: simply its animus is satisfied with less overt fuss. Algorithms don't have emotions. So there is plenty of innovative risk taking, but low volatility by historical standards, induced by trading technologies, has not yet been properly internalized by many market participants. Viewing contemporary volatility levels in the manner to which historical experience has been accustomed ineluctably leads to excessive risk taking.

Chapter 10 is interesting in its own right, notwithstanding any relationship to the evolution of statistical arbitrage opportunities. Algorithms and computer driven trading are changing the financial world in many ways. Electronic exchanges have already been seen off most of the world's peopled trading places—and who among us believes that the floor of the NYSE will be more than a museum, parking lot, or memory in a year or two?

Chapter 11 describes the phoenix of statistical arbitrage, rising out of the ashes of the fire created and sustained by the technological developments in algorithmic trading. New, sustained patterns of stock price dynamics are emerging. The story of statistical arbitrage has returned to a new beginning. Will this fledgling fly?

The renaissance predicted in Chapter 11, drafted in 2005, is already coming to pass. Since at least early 2006 there has been a resurgence of performance from those practitioners who persisted through the extremely challenging dynamic changes of 2003–2005. Interestingly, while there are new systematic patterns in the movements of relative equity prices, some old patterns have also regained potency. Adoption of algorithmic trading is accelerating, with tools now offered by more than 20 vendors. In another technology driven development, beginning with Goldman Sachs in late 2006, at least two offerings of general hedge fund replication by algorithmic means have been brought to market. This is an exciting as well as exacting time for statistical arbitrageurs.

Foreword

Mean reversion in prices, as in much of human activity, is a powerful and fundamental force, driving systems and markets to homeostatic relationships. Starting in the early 1980s, statistical arbitrage was a formal and successful attempt to model this behavior in the pursuit of profit. Understanding the arithmetic of statistical arbitrage (sometimes abbreviated as stat. arb.) is a cornerstone to understanding the development of what has come to be known as complex financial engineering and risk modeling.

The trading strategy referred to as statistical arbitrage is generally regarded as an opaque investment discipline. The view is that it is being driven by two complementary forces, both deriving from the core nature of the discipline: the vagueness of practitioners and the lack of quantitative knowledge on the part of investors. Statistical arbitrage exploits mathematical models to generate returns from systematic movements in securities prices. Granted, no investment manager is inclined to divulge the intricate "how-tos" of his business. While stock pickers can tell a good story without revealing the heart of their decision making, that is not the case with model-based strategies developed by "quants." A description with any meaningful detail at all quickly points to a series of experiments from which an alert listener can try to reverse-engineer the strategy. That is why quant practitioners talk in generalities that are only understandable by the mathematically trained.

Opacity has also increased the need for mathematical maturity on the part of investors seeking to assess managers. To comprehend what a statistical arbitrageur is saying beyond a glib level, one needs to understand advanced mathematics beyond the college level. This, naturally, limits the audience. The limitation is perpetuated by the lack of reference material from which to learn. *Statistical Arbitrage* now fills that void.

Statistical arbitrage has been in existence for approximately 25 years. During that time, the general concepts have been widely disseminated via the storytelling of early implementers to interested investment bank analysts and academics. Nevertheless, opacity remains because practitioners have steadily increased the sophistication of their modeling—and for good commercial reasons remained obscure about their innovations. In the wide dissemination of basic stat. arb. concepts, the term mean reversion as well as its variant, reversion to the mean, looms very large. Reversion to the mean is a simple concept to illustrate: Children of unusually tall parents are typically shorter than their parents; children of unusually short parents are typically taller than their parents. This is a concept that is easy for most people to grasp. Translating this idea to the motions of security prices means that securities prices return to an average value. So far, so good. But then we hit a problem. Height reversion is an intergenerational phenomenon, while price reversion is an entity dynamic.

Prices returning from where? And to what average value? The average height of adults is a familiar concept, even if the precise quantification requires a little work. Even children as young as grade-school age can give a reasonable estimate of the average height of the adults they know, and by extension, of the average height of local adult populations. There is no such common grounding of observation or experience to apply to securities prices. They are all over the map. Scaling is arbitrary. They can grow many fold. And they can collapse to zero. People do not grow to the sky and then revert back to some average, but security prices can.

Even if we suppose that the questions have been reasonably answered, other technicalities immediately pose themselves: How does one identify when a price is away from the mean and by how much? How long will the return to the mean take?

Here is where the opacity enters the discussion and makes its permanent home. The language of mathematical models compounds the unfamiliarity of the notions, generating a sense of disquiet, a fear of lack of understanding.

In *Statistical Arbitrage,* Pole has given his audience a didactic tour of the basic principles of statistical arbitrage, eliminating opacity at the Statistical Arbitrage 101 level. In the 1980s and early 1990s, Stat. Arb. 101 was, for the most part, all there was (exceptions such as D.E. Shaw and Renaissance aside). Today, more than a decade later, there is a much more extensive and complex world of statistical arbitrage.

This is not unlike the natural world, which is now populated by incredibly complex biological organisms after four billion years of evolution. Yet the simplest organisms thrive everywhere and still make up by far the largest part of the planet's biomass. So is it true in statistical arbitrage, where the basics underpin much of contemporary practice.

Statistical Arbitrage describes the phenomena, the driving forces generating those phenomena, the patterns of dynamic development of exploitable opportunities, and models for exploitation of the basic reversion to the mean in securities prices. It also offers a good deal more, from hints at more sophisticated models to valuable practical advice on model building and performance monitoring—advice applicable far beyond statistical arbitrage.

Chapters 1 and 2 speak to the genesis of statistical arbitrage, the venerable pairs trading schemes of the 1980s, with startling illustration of the enormous extent and productivity of the opportunities. This demonstration sets the scene for theoretical development, providing the first step to critical understanding of practical exploitation with rules for calibrating trade placement. More penetration of opacity follows in Chapter 5 where the relationship between (a) reversion in securities prices watched day-by-day and (b) statistical descriptions (distributions) of collections of such daily prices viewed as a glob devoid of the day-by-day context, is clearly spelled out.

Chapters 8 and 9 tell of the midlife crisis of statistical arbitrage. The roiling of United States financial markets for many months, beginning with the Enron debacle in 2000 and running through the terrorist attacks of 2001 and what Pole calls "an appalling litany" of corporate misconduct, is dissected for anticipated impact on statistical arbitrage performance. Adding to that mix have been technical changes in the markets, including decimalization and the decline of independent specialists on the floor of the NYSE. Pole draws a clear picture of why statistical arbitrage performance was disrupted. Very clearly the impression is made that the disruption was not terminal.

Chapters 10 and 11 speak to the arriving future of statistical arbitrage. Trading algorithms, at first destroyers of classical stat. arb. are now, Pole argues, progenitors of new, systematically exploitable opportunities. He labels one of the new motions the "catastrophe move"; a detailed exposition of modeling the dynamics follows a catastrophe-theory explication of a possible rationale for the behavioral pattern. The unmistakable impression is that statistical arbitrage is rising once again.

The tone of *Statistical Arbitrage* is direct and thorough. Obfuscation is in short supply. Occasionally, the tone is relieved with a bit of lightheartedness—the tadpole-naming story in a note to Chapter 11 is a gem—and throughout, refreshing prose is to be found.

In describing mathematical models, authors readily produce unmemorable, formulaic wording offering nothing by way of interpretation or explanation beyond what is provided by the algebra itself. *Statistical Arbitrage* is an exception—a break in the cloud of opacity—a mean that Pole has avoided reverting to!

Gregory van Kipnis

April 23, 2007
New York City

Acknowledgments

I was introduced to statistical arbitrage by Gregg van Kipnis. In many ways, the contents of this volume are directly the result of our collaboration and it is a pleasure to acknowledge the intellectual debt. Our conversations often strayed far beyond statistical arbitrage to macroeconomics and general science and very often to politics, none of which is reproduced here in recognizable form. Those discussions were not always motivated by statistical arbitrage considerations, though occasionally we would hit on a useful metaphor from an unrelated topic that subsequently proved fruitful in thinking about statistical arbitrage. It is not in the nature of things that individual such recollections can now be pointed to with certainty to say whose idea ABC was. Credit is rightfully due to van Kipnis; the rendition in these pages is entirely my responsibility.

The editorial and production staff at Wiley, in particular Bill Falloon, Emilie Herman, Laura Walsh, and Stacey Fischkelta though we never physically met, have been helpful and courteous throughout the project.

CHAPTER 1

Monte Carlo or Bust

We must always be ready to learn from repeatable occurrences however odd they may look at first sight.
—*Box on Quality and Design*, G.E.P. Box

1.1 BEGINNING

In 1985 a small group of quantitatively trained researchers under the tutelage of Nunzio Tartaglia[1] created a program to buy and sell stocks in pair combinations. Morgan Stanley's Black Box was born and quickly earned a reputation and a lot of money. A fifteen-year rise to heroic status for statistical arbitrage (a term uncoined at that time) was begun.

Details of the Black Box were guarded but soon rumor revealed the basic tenets and the name "pairs trading" appeared in the financial lexicon. The premise of pairs trading was blindingly simple: Find a pair of stocks that exhibit similar historical price behavior. When the prices of the stocks diverge, bet on subsequent convergence. Blindingly, beautifully simple. And hugely profitable.

[1]In *The Best of Wilmott*, Paul Wilmott states that the MS pairs trading program was initiated by Gerry Bamberger in 1982/1983, that Bamberger departed MS in 1985 for Princeton Newport Partners and retired in 1987. We are unable to confirm whether Bamberger's MS program was distinct from Tartaglia's; others have claimed a group effort and complain that it is unfair to annoint either group head as "the inventor."

Interestingly Wilmott claims that pairs trading was discovered at his firm as early as 1980.

FIGURE 1.1 Daily closing prices, CAL and AMR (2002–2004)

Where did Tartaglia get his insight? As with many stories of invention, necessity was the motivating force. Chartered by management to find a way to hedge the risks routinely incurred through its lucrative activities with block trading, Tartaglia's mathematical training conjured up the notion of selling (short) a stock that exhibited similar trading behavior to the stock being managed by the block desk. Immediately the notion was invented, the more general application of pairs trading was innovated. Very shortly, a new profit center was adding to the bottom line.

Figure 1.1 shows the daily close price of two airline stocks, Continental Airlines (CAL) and American Airlines (AMR). Notice how the spread between the two price traces opens and closes. The pairs trading scheme veritably yells at one: Buy the lower-priced stock and short the higher-priced stock when the spread is "wide" (A), and reverse out those positions when the spread closes (B).

In 1985 computers were not familiar appliances in homes, and daily stock price feeds were the tools of professionals alone. Sheer number crunching power, crucial to serious implementation of a pairs trading business, required tens of thousands of dollars of hardware. Pairs trading, so beautifully simple in concept and for years now in practice, was born in an era in which investment houses alone could realistically research and deploy it.

Many stories from the era infect the industry, mythologizing the business and the practitioners. Two such stories that have genuine substance and that have continued significance today are the SEC's use of algorithms to detect aberrant price patterns, and the evolution of specialist reaction to the contrarian traders from initial suspicion to eventual embrace.

The SEC was intrigued no less than others by the aura around the Morgan Stanley black box. Upon learning about how the models worked to predict certain stock price motions, it was quickly realized how the technology could be employed to flag some kinds of unusual and potentially illegal price movement, long before neural network technology was employed in this role.

In the late 1980s the NYSE was populated with over 50 independent specialists. Largely family businesses with limited capital, they were highly suspicious when the group at Morgan Stanley began systematically sending orders to "buy weakness" and "sell strength." The greatest concern was that the big house was attempting to game the little specialist. Suspicion gradually evolved into cozy comfort as the pattern of trading a stock was revealed. Eventually, comfort became full embrace such that when the specialist saw Morgan Stanley accumulating a weak stock, the specialist would jump on the bandwagon "knowing" that the stock's price was set to rise.

The early years were enormously lucrative. Success soon spawned independent practitioners including D.E. Shaw and Double Alpha, both created by former acolytes of Tartaglia. In subsequent years other groups created pairs trading businesses, the founders of which can be traced either to the original group at Morgan Stanley or to second-generation shops such as Shaw. As the practice became more widely known, academic interest was piqued; published articles by NBER, among others, made the general precept known to a wide audience and with the rapid increase in power of low cost personal computers, the potential practitioner base exploded. Very quickly, so did the actual practitioner base.

1.2 WHITHER? AND ALLUSIONS

Two decades later, the matured adult statistical arbitrage that grew up from the pair trading infant faces a cataclysmic environmental change. Returns have greatly diminished. Managers are beset by difficulties and are adapting strategies to cope. The financial market environment of the new century poses survival challenges one might liken to those faced by earthly fauna millenia ago when the last ice age dawned. The quick and adaptable survived. The slow and morphologically fixed froze or starved.

Statistical arbitrage's ice age dawned in 2000 and entered full "frigidia" in 2004. Observers proclaimed the investment discipline's demise, investors withdrew funds, and practitioners closed shop. The rout was comprehensive. A pall of defeat enveloped discussion of the business.

This judgment of a terminal moment for statistical arbitrage is premature, I believe. Despite the problems for traditional statistical arbitrage models presented by market structural changes, which are documented and examined in later chapters, there are indications of new opportunities. New patterns of stock price behavior are occurring on at least two high-frequency timescales. Driving forces are identifiable in the interplay of electronic trading entities, the rising future of stock trading in the United States.

The appearance of the new opportunities, admittedly only roughly characterized at this time, suggests significant economic exploitability, and they may be enough to stave off the fate of extinction for statistical arbitrage. The cro magnon man of classic reversion plays will be superseded by the homo sapiens of.... That remains to be seen but outlines are drawn in Chapter 11.

I considered titling the book, *The Rise and Fall and Rise? of Statistical Arbitrage*, reflecting the history and the possibilities now emerging. The pattern is explicit in the preceding paragraphs of this chapter and in the structure of the book, which is written almost in the form of an annotated history. To those readers whose interest is borne of the question, "What are the prospects for statistical arbitrage?", the historical setting and theoretical development in Chapters 1 through 7 may seem anachronistic, unworthy of attention. It might be likened to suggesting to a student of applied mathematics that the study of Copernicus' system for the motions of astronomical bodies is presently utilitarian. I maintain that there is value in the historical study (for the mathematician, too, but that is taking analogy much further than it deserves). Knowing what worked previously in statistical arbitrage, and how and why it did, provides the necessary foundation for understanding why market structural changes have negatively impacted the strategy class. Knowing which changes have had an effect and how those effects were realized illuminates what might be anticipated in the presently congealing environment.

Interpreting the present in the context of the past is hardly a novel notion. It is a sound bedrock of scientific investigation. Most people are familiar with the admonition of political philosophers that those who do not study the past are doomed to repeat its mistakes.[2] But that is not our reference point. While undoubtedly some arbitrageurs have made their individual errors, there cannot be a verdict that the collective of practitioners has "made a mistake" that ought to be guarded against ever after. Our reference point is the far more compelling scientific view of "standing on the shoulders of giants." Bereft of value judgments, scientific theories, right or wrong, and no matter how pygmy the contribution, are set forth for scrutiny forever. The promise of the new opportunities may be understood and evaluated in the context of how market changes rendered valueless that which was formerly lucrative.

Let's be quite clear. There is no claim to a place in history with the work reported here despite allusions to historical scientific genius. Neither is the area of study justifiably on the same shelf as physics, chemistry, and mathematics. It sits more appropriately with economics and sociology because the primal forces are people. We may label an emergent process as "reversion" (in prices), describe temporal patterns, posit mathematical equations to succinctly represent those patterns, and commit ourselves to actions—trading—on

[2]"Progress, far from consisting in change, depends on retentiveness. When change is absolute there remains no being to improve and no direction is set for possible improvement: and when experience is not retained, as among savages, infancy is perpetual. Those who cannot remember the past are condemned to repeat it. In the first stage of life the mind is frivolous and easily distracted, it misses progress by failing in consecutiveness and persistence. This is the condition of children and barbarians, in which instinct has learned nothing from experience." *The Life of Reason*, George Santayana.

the output of the same, but the theory, models, and analysis are of an emergent process, not the causal mechanism(s) proper. No matter how impressibly we may describe routines and procedures of the regular players, from analysts (writing their reports) to fund advisors (reading those reports, recommending portfolio changes) to fund managers (making portfolio decisions) to traders (acting on those decisions), the modeling is necessarily once removed from the elemental processes. In that complex universe of interactions, only the result of which is modeled, lies the genesis of the business and now, more fatefully, the rotting root of the fall. Astonishingly, that rotting root is fertilizing the seeds of the rise(?) to be described.

Unlike the study of history or political philosophy, which is necessarily imbued with personal interpretations that change with the discovery of new artifacts or by doubt cast on the authenticity of previously sacred documents, the study of statistical arbitrage benefits from an unalterable, unequivocal, complete data history that any scholar may access. The history of security prices is, like Brahe's celestial observations, fixed. While Brahe's tabulations are subject to the physical limitations of his time[3] and uncertainties inherent in current relativistic understanding of nature's physical reality, the history of security prices, being a human construct, is known precisely.

In exhorting the quality of our data, remember that Brahe was measuring the effects of physical reality on the cosmic scale for which scientific theories can be adduced and deduced. Our numbers, records of financial transactions, might be devoid of error but they are measurements of bargains struck between humans. What unchanging physical reality might be appealed to in that? We might build models of price changes but the science is softening as we do so. The data never changes but neither will it be repeated. How does one scientifically validate a theory under those conditions?

---

[3]The first Astronomer Royal, John Flamsteed (1646–1719), systematically mapped the observable heavens from the newly established Royal Observatory at Greenwich, compiling 30,000 individual observations, each recorded and confirmed over 40 years of dedicated nightly effort. "The completed star catalogue tripled the number of entries in the sky atlas Tyco Brahe had compiled at Uraniborg in Denmark, and improved the precision of the census by several orders of magnitude." In *Longitude* by Dava Sobel.

The questions are unanswerable here. One cannot offer a philosophy or sociology of finance. But one can strive for scientific rigor in data analysis, hypothesis positing, model building, and testing. That rigor is the basis of any belief one can claim for the validity of understanding and coherent actions in exploiting emergent properties of components of the financial emporium.

This volume presents a critical analysis of what statistical arbitrage is—a formal theoretical underpinning for the existence of opportunities and quantification thereof, and an explication of the enormous shifts in the structure of the U.S. economy reflected in financial markets with specific attention on the dramatic consequences for arbitrage possibilities.

CHAPTER 2

Statistical Arbitrage

Much of what happens can conveniently be thought of as random variation, but sometimes hidden within the variation are important signals that could warn us of problems or alert us to opportunities.

—*Box on Quality and Discovery*, G.E.P. Box

2.1 INTRODUCTION

The pair trading scheme was elaborated in several directions beginning with research pursued in Tartaglia's group. As the analysis techniques used became more sophisticated and the models deployed more technical, so the sobriquet by which the discipline became known was elaborated. The term "statistical arbitrage" was first used in the early 1990s.

Statistical arbitrage approaches range from the vanilla pairs trading scheme of old to sophisticated, dynamic, nonlinear models employing techniques including neural networks, wavelets, fractals— just about any pattern matching technology from statistics, physics, and mathematics has been tried, tested, and in a lot of cases, abandoned.

Later developments combined trading experience, further empirical observation, experimental analysis, and theoretical insight from engineering and physics (fields as diverse as high energy particle physics to fluid dynamics and employing mathematical techniques from probability theory to differential and difference equations). With so much intellectual energy active in research, the label "pairs trading" seemed inadequate. Too mundane. Dowdy, even. "Statistical arbitrage" was invented, curiously, despite the lack of statisticians or statistical content of much of the work.

2.2 NOISE MODELS

The first rules divined for trading pairs were plain mathematical expressions of the description of the visual appearance of the spread. For a spread like the CAL–AMR spread in Figure 2.1, which ranges from −\$2 to \$6, a simple, effective rule is to enter the spread bet when the spread is \$4 and unwind the bet when it is \$0.

We deliberately use the term *rules* rather than *model* because there is no attempt at elaboration of a process to explain the observed behavior, but simply a description of salient patterns. That is not to diminish the validity of the rules but to characterize the early work accurately. As the record shows, the rules were fantastically profitable for several years.

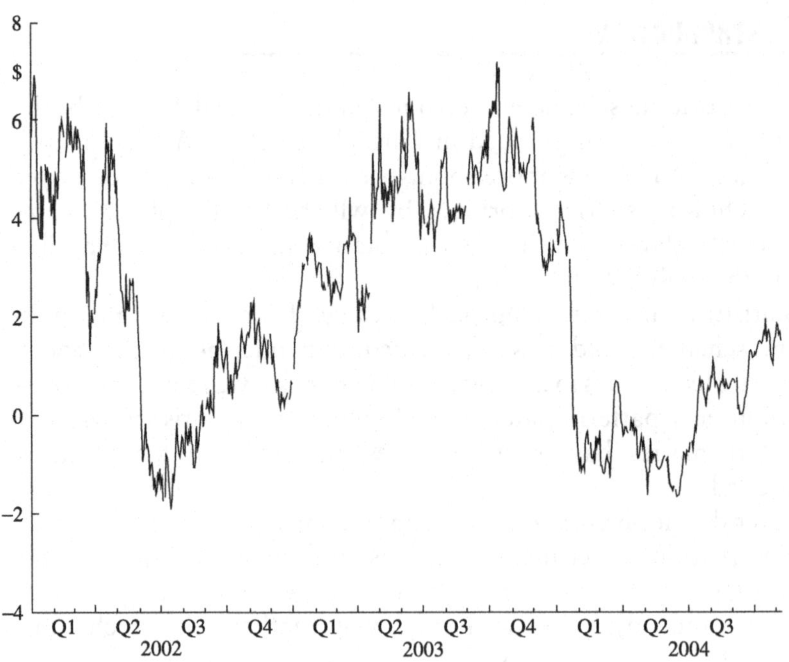

FIGURE 2.1 Daily closing spread, CAL–AMR

Applying the \$4–\$0 rule to the CAL–AMR spread, there is a single trade in the calendar years 2002 and 2003. If this looks like money for practically no effort, that is the astonishing situation Tartaglia discovered in 1985—writ large across thousands of stock pairs.

Alternatives, elaborations, and generalizations jump off the page as one looks at the spread and considers that first, seductively simple rule. Two such elaborations are:

- Make the *reverse* bet, too.
- Make repeated bets at staged entry points.

2.2.1 Reverse Bets

Why sit out the second half of 2002 while the spread is increasing from its narrow point toward the identified entry point of \$4? Why not bet on that movement? In a variant of the commodity traders' "turtle trade," rule 1 was quickly replaced with rule 2, which replaced the *exit* condition, "unwind the bet when the spread is \$0," with a *reversal*, "reverse the long and short positions." Now a position was always held, waiting on the spread to increase from a low value or to decline from a high value.

With that expansion of trading opportunities came more trades and greater profits for no additional work.

2.2.2 Multiple Bets

In the first quarter of 2002 the CAL–AMR spread varies over a \$6 range from a high of \$7 to a low of \$1. Bets placed according to rule 1 (and rule 2) experience substantial mark to market gains and losses but do not capture any of that commotion. Since the spread increases and decreases over days and weeks, meandering around the trend that eventually leads to shrinkage to zero and bet exit (rule 1) or reversal (rule 2), why not try to capture some of that movement?

Rule 3 is designed to extract more from spreads by adding a second entry point to that identified in rule 1. For CAL–AMR the rule is: Make a second bet on the subsequent shrinking of the spread when the spread increases to \$6. Doubled bets on the spread shrinkage would be made in both 2002 and 2003, increasing profit

FIGURE 2.2 Daily closing prices, CAL and AMR (2000)

by 150 percent! (Profit is increased by a smaller percentage in 2002 over that obtained with rule 2 because rule 2 gains from the reverse bet which is unaltered in rule 3. There was no reverse bet in 2003, the position being carried into 2004.)

This single illustration demonstrates in blinding clarity the massive opportunity that lay before Tartaglia's group in 1985, an era when spreads routinely varied over an even wider range than exhibited in the examples in this chapter.

2.2.3 Rule Calibration

Immediately when one extends the analysis beyond a single pair, or examines a longer history of a single pair, the problem of calibration is encountered. In Figure 2.2 another pair of price histories is shown,

FIGURE 2.3 Daily closing spread, CAL–AMR

now for the single year 2000. Figure 2.3 shows the corresponding spread.[1]

Wow! We should have shown that example earlier. The spread varies over a $20 range, three times the opportunity of the CAL–AMR

[1]The price series for AMR is adjusted for the spinoff of Sabre, the company's reservations business, on March 16, 2000. Without proper adjustment, the close price series would drop from $60 to $30 overnight—an unrealistically dramatic spread change! We elected to adjust prices moving back in time, so that the pre-spinoff prices are altered from the values that obtained in the market at the time, preserving more recent prices. Trading AMR in January 2000, one would of course have been working at the actual pre-spinoff level of circa $60. How one makes price adjustments, forward or backward, is a matter of taste, though it must be done consistently. Return series computed from adjusted price histories are unique and for that and other reasons, most analysis is done in terms of returns rather than prices. In this book, prices are used for demonstration because the elucidated points are more graphically made therewith. Price adjustment for corporate events, including dividends and splits, is critical to proper calculation of gains from trading.

example examined in Figure 2.1. But right there in that rich opportunity lies the first difficulty for Rules 1–3: The previously derived calibration is useless here. Applying it would create two trades for Rule 3, entering when the spread exceeded $4 and $6 in January. Significant stress would quickly ensue as the spread increased to over $20 by July. Losses would still be on the books at the end of the year. Clearly we will have to determine a different calibration for any of Rules 1–3. Equally clearly, the basic form of the rules will work just fine.

Now consider the problem of calibration applied to hundreds or thousands of potential spreads. Eyeballing graphs would require a lot of eyeballs. A numerical procedure, an automatic way of calibrating rules, is needed. Enter statistics. The trading rules were divined by visually determining the range over which the spread varied. This is trivially computed automatically: The maximum and minimum spread in Figure 2.1 is −$2 and $7. Allowing a margin of, say, 20 percent, an automatic calibration would give entry and exit values of $5 and $0 for rule 1. This is not exactly what we selected by eye, but operationally it generates similar (though richer) trades. Critically, the procedure is readily repeated on any number of spreads by computer.

For the second example (Figure 2.2) the spread range is $3 to $22. The 20 percent margin calibration gives trade entry and exit values of $18 and $7 respectively. Applying Rule 1 with this automatic calibration yields a profitable trade in 2000. That desirable outcome stands in stark contrast to the silly application of the example one calibration (entry at $4 and $6 and unwind at $0 as eyeballed from Figure 2.1) to the spread in Figure 2.2 which leads to nauseating mark to market loss.

Calibration Epochs In the foregoing discussion of eyeball calibration, we did not make explicit the span of time being considered, which is two years in the first example, one in the second. Naturally, both examples were selected to convey in stark terms the beautiful simplicity and evident availability of the pair trading opportunity. Nevertheless, the examples are not unrealistic. And so: How much time is appropriate?

The stocks in Figures 2.1 and 2.2 are the same: CAL and AMR. The question, "How much time is appropriate?", is now seen to be dramatically important continually, not just as a once-only decision for each candidate pair. Imagine the consequences of using the calibration of the CAL–AMR spread from 2000 for trading in 2002–2003. In this case, the consequences look benign: no trades. But that is a negative consequence because valuable trading opportunities are missed. In other cases, horribly costly bets would be placed using a rule calibrated on out-of-date price history.

This question of how much price history to use to calibrate a trading rule is critical. In contrast to the analysis described thus far, one shot, static analysis in which the rule is applied to the same price history as that from which it was derived, practical trading is always an application of the past to the unknown future. In Figure 2.4, the four-year spread (2000–2003) history for CAL–AMR is shown together with upper and lower limits, maximum −20 percent range and minimum +20 percent range, respectively, calculated with a look back window of three months. While these limits are, at times, not nearly as good as the eyeball limits previously examined, they do

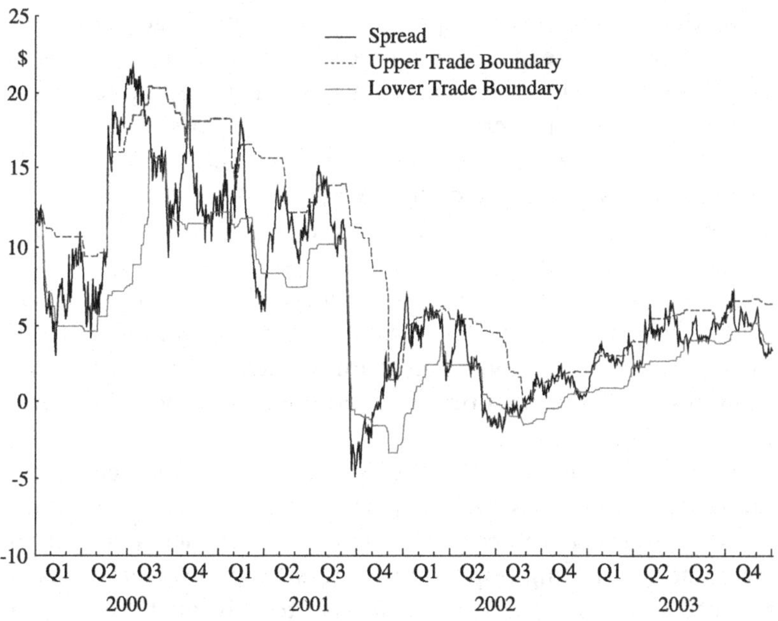

FIGURE 2.4 Daily closing spread, CAL–AMR (with upper and lower trade rule margins)

retain good properties for trade identification. Furthermore, unlike the previous *in sample* calculations, the current estimates are *out of sample* projections. On any day, the only price information used in the calculation is publicly available history. The computed limits are, therefore, practically actionable.

Applying Rule 2, there are 19 trades (ignoring the first quarter of 2000 since the limits are computed on insufficient data), comprising 4 losing trades and 15 winning trades. Both winning and losing trades exhibit periods where substantial mark to market losses are incurred before gains accrue toward the end of the trade. One last observation: Notice how the volatility of the spread has substantially declined from 2000–2003; much will be said about that development in later chapters.

2.2.4 Spread Margins for Trade Rules

In response to the demonstrated problem of determining operational limits on the spread range to guide trade decisions, we chose to use margins of 20 percent. In the three-month window the upper boundary, "short the spread," is *max* spread −20 percent range, the lower boundary, "buy the spread," is *min* spread +20 percent range. This operational procedure has the great merit of ready interpretation. It is unambiguously clear what the margins are: one fifth of the calculated range of the spread over the previous three months.

Less satisfactory is the use of the extreme values, *max* and *min*. Extremes exhibit great variability. Projecting extremes is therefore subject to great uncertainty: Think of outliers and all you have read in statistics texts about careful analysis thereof. Modeling extremes is a complicated and fascinating area of study with applications ranging from peak river flow for flood prediction to electricity demand for prediction of generation requirements and the likelihood of outages, among many others.

From the extensive variability of extremes comes the need for a substantial margin (20 percent) for practicable trade rules. Suppose that just the largest 10 percent of spread displacements served to generate sufficient trading opportunities to sustain a business based on historical analysis. It would be unwise to rely on that margin for actual trading because of the inherent uncertainty in projecting information into the future. Extremes in the future are certain to be different from extremes in the past. If a spread experiences a "quiet period," business will be poor because few opportunities will be identified even though there may be plenty of profitable opportunities. Better to be conservative and use a larger margin. Of course, a spread may exhibit a volatile period; the consequences there are more volatility in mark to market revenue but not a reduction in business or total profit.

Greater stability is obtained between extremes. Projecting the central location of the spread is done with considerably greater confidence than projecting the extremes. Therefore, most implementations modify the "go short" and "go long" limits to be computed as offsets from the center rather than offsets from the extremes. Bollinger bands, mean plus or minus a standard deviation, are a classic example. In spite of the rationale though, it is arguable how much stability is improved by this switch of focus: The standard deviation is computed from all the data, extremes included, and since observations are squared, the extreme values actually receive proportionally greater weight! Robust procedures are sensibly employed, which amounts to excluding (in more sophisticated applications, down-weighting) the most extreme values in a sample before computing summary statistics such as the mean and standard deviation.

Extrapolating percentage points of the spread distribution, say the twentieth and eightieth percentile, is similarly robust but is seldom seen. Operationally it is of no practical significance in the simple trading rules described here. Greater significance is found where models are more sophisticated and the asymmetry of distributions has mercenary implications.

Greater (presumed) stability is achieved at the cost of some interpretability. There is no unique relationship between the standard deviation of a distribution and the range. When presented with standard deviations, many assume or don't realize they are assuming an underlying normal distribution and equate mean plus and minus one standard deviation as two-thirds probability and mean plus and minus two standard deviations as 95 percent probability. Financial data is typically nonnormal, exhibiting asymmetry and significantly more observations several standard deviations from the mean, the so-called "heavy tails." These tails are heavy only by comparison to the normal distribution, not by what is typical in finance data. The use of tail area probabilities from the normal distribution is therefore a frequent cause of miscalculation—and usually that means underestimation—of risk. Most errors of this kind are trivially avoided by using the empirical distribution—the data itself—rather than assumed mathematical forms. Moreover, it is quite simple to examine the fit of a normal curve to a set of data and judge the accuracy of probability calculations for intervals of interest, be they in the tail or center of the distribution. Chapter 5 demonstrates these points in a discussion of reversion in price series.

With so many potentially costly errors attached to the use of sample moments (mean and standard deviation) why is the range so readily abandoned? What has been gained by the sophistry? In addition to the aforesaid (almost unconscious) action on the part of many, there is the conscious action on the part of many others that is driven by mathematical tractability of models. Extreme values (and functions thereof) are difficult to work with analytically, whereas standard deviations are generally much easier. For the normal distribution typically assumed, the mean and standard deviation are defining characteristics and are therefore essential.

While the technicalities are important for understanding and analysis, the practical value for application in the late 1980s and early 1990s was minimal: Reversion was evident on such a large scale and over such a wide range of stocks that it was impossible not to make good returns except by deliberate bad practice! That rich environment has not existed for several years. As volatility in some industries declined—the utilities sector is a splendid example (Gatev, et al.)—raw standard deviation rules were rendered inadequate as the expected rate of return on a trade shrank below transaction costs. Implementing a minimum rate of return lower bound on trades solved that, and in later years provided a valuable risk management tool.

2.3 POPCORN PROCESS

The trading rules exhibited thus far make the strong statement that a spread will systematically vary from substantially above the mean to substantially below the mean and so forth. The archetype of this pattern of temporal development is the sine wave. In the early years of pairs trading, that archetype provided the theoretical model for

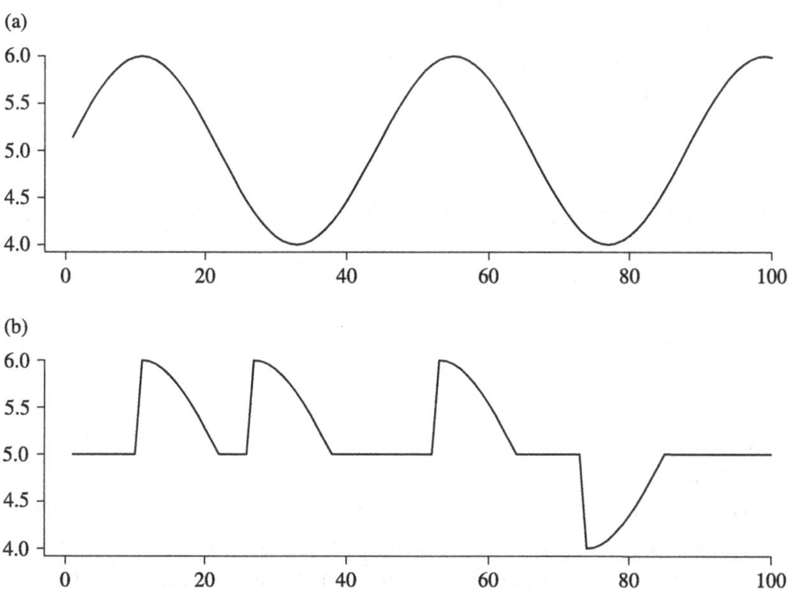

FIGURE 2.5 Process archetypes: (a) sinusoidal, (b) popcorn

spread analysis, but many trade opportunities were observed to be missed. An alternative archetype, which we shall call the "popcorn process," shown in Figure 2.5, provided new insight. Reversion to the mean following a disturbance from that mean was more closely focused upon. In this model, the constraint on spread motion of undulation (even if more irregularly than the mathematical archetype) is removed. An upward motion (move to a "distant" peak) may be followed, after return to the local mean, by another excursion to a distant peak. Similarly a decline to a distant trough may follow a previous excursion to a distant trough without an intervening move to a distant peak. The qualifier "distant" is used here to distinguish substantive moves from the mean from minor variation about the mean. Two troughs are by definition separated by a peak but a peak is of trading interest only if it is sufficiently removed from the mean such that movement back to the mean is economically beneficial. The important point here is that a peak separating troughs can be near the mean and is not forced or assumed to be substantially above the mean.

Expressing the popcorn process mathematically is more complicated than writing a sine function, but not much so. If the sine function is written:

$$y_t = \sin(t)$$

then the popcorn function may be similarly written:

$$y_t = I_t \sin(t)$$

where I_t is an indicator function taking values 1 or -1 signaling a peak or a trough move. The math is not important here; the insight from the description and graphical depiction of the process is: Exploiting the popcorn process is not efficiently accomplished using the turtle trade. In Figure 2.5, panel (b), the turtle trade rule identifies a single trade with profit $2. The popcorn process suggests a rule that signals to exit a trade when the spread returns to the mean, rather than assuming it will continue beyond the mean to an extreme on the opposite side from which the trade was entered. This new rule identifies four trades with total profit $4. Another novel feature of the rule is that it contains periods where no capital is committed.

All the necessary calculations for the new rule have already been described: local mean and range of the spread. The change is to the trade rule.

Rule 4: When the spread increases (decreases) sufficiently far from the mean (say, *k* standard deviations) sell (buy) the spread; unwind the positions when the spread returns to the mean.

Many of the more elaborate models built by statistical arbitrageurs, whether for pairwise spreads or more complicated functions of stock price histories, are based on the understanding of the popcorn process, or reversion to the mean, rather than the sinusoidal or turtle trade process. Chapter 3 describes some of the models and modeling considerations. The interesting phenomenon of stochastic resonance (also described in Chapter 3) admits a valuable modification of the exit condition in Rule 4.

2.4 IDENTIFYING PAIRS

The opportunity is huge. We have a set of operational trading rules and automatic calibration procedures. Now, which pairs can we trade?

Early on, stocks were grouped by broad industry classification and every pair within those groups was a candidate. Risk management was rudimentary with Barra models applied to constructed portfolios and identified factor exposures offset by trades away from the pair portfolio (using stocks or Standard and Poor's (S&P) futures to neutralize β exposure, for example).

Elaborations were introduced as greater control over return variability became desirable and as experience showed where structural weaknesses lay. Individual manager preference became influential when hedge funds began marketing pairs trading and statistical arbitrage strategies.

Maximizing correlations was an early filter applied to pair selection: Compute the correlation of each candidate pair (using, for example, two years of daily data) and retain only those pairs having correlations greater than some minimum. On the assumption that past correlation is predictive of future correlation, this filtering eliminates pairs of stocks that exhibit little or no relationship. The rationale holds that uncorrelated stocks are behaviorally unrelated and, hence, unpredictable as a pair.

2.4.1 Refining Pair Selection

Reversion betting on pair spreads works best when the two constituent stock prices continually move apart and together again. That pattern of behavior, stock A increasing when stock B decreases and vice versa, generates very *low* (even negative) correlation. So from a profit or return perspective, were the early correlation filters (searching for a high degree of correlation) quite wrong? No: In the short term, profits may be forgone by excluding low correlation pairs but the long-run risk situation is greatly improved. Stocks that typically exhibit contrary or unrelated price movements are more likely to respond disparately to fundamental market developments than stocks that tend to move together. At some point, unrelated stocks are very likely to create a costly pair trade.

That insight motivates a subtly different approach to the correlation filter. Defining risk moments (or events) as times when a stock price trace changes direction such that a peak or trough is formed, it is desirable for risk minimization purposes to select pairs that show similar event histories—peaks and troughs close in time with similar sized moves for the two stocks between these events. Such pairs are less likely to react divergently (except, perhaps, in the immediate aftermath) following a disturbance to the market (political, industrial development, etc.). For profit maximization, it is desirable that between events the two stocks develop along different price trajectories, exhibiting as much negative correlation—moving apart, then together—as possible. See Chapter 5 for a formal treatment of desirable and undesirable pair correlations.

2.4.2 Event Analysis

The turning point algorithm works as follows:

1. A local maximum in the price series is a turning point if subsequently the price series declines by an amount giving a negative return greater in absolute value than a specified fraction of the local, annualized return volatility.
2. Similarly, a local price minimum is a turning point if subsequently the price rises by an amount giving a return greater than the specified fraction of local, annualized return volatility.
3. Look at the price trace in Figure 2.6 (General Motors, daily adjusted prices). Given a turning point identified at a, where is the next turning point? Point a is clearly a local minimum; therefore, the next turning point must be a local price maximum. Move forward in time looking at the price series from a to t. Identify the local maximum price in the interval $[a, t]$; call it p. Is the decline from the price at p to the price at t greater than k percent of the local volatility at t (looking back)?
4. When $p = m$ and $t = t_1$, the answer is no. Not until b is identified as the local maximum ($t > t_2$) and then not until $t = t_3$, is the answer yes.
5. For this example, specification is for a window of 20 days to define local volatility, an annualization factor of 16, and a turning point qualifying fraction of 30 percent.

Figure 2.7 shows the General Motors price series again, this time with turning points identified with a less demanding criterion: A decline in price from a peak by 25 percent of the local volatility qualifies the peak as a turning point. Four additional local extrema are

FIGURE 2.6 Adjusted close price trace (General Motors) with 30 percent turning points identified

identified (ignoring the series end point) compared with the default 30 percent criterion. Still, two local peaks and troughs in mid-1997 are not identified by the algorithm. They provide returns of about −4 percent in a few days—a fabulous annualized rate of return.

Figure 2.8 shows the General Motors price series once more, with an even less stringent turning point criterion: A decline in price from a peak by 20 percent of the local volatility qualifies the peak as a turning point. Eight additional local extrema are identified (ignoring the series end point) compared with the default 30 percent criterion, the same additional four identified by the 25 percent criterion plus another four.

In other examples, changing the window length, the strict capture by a less stringent criterion of the complete set of turning points identified by a more stringent criterion is not observed. These examples and observations serve as reminders that the analysis here is strictly statistical. The events reflect market sentiment but that may be driven by news unrelated to the stock or by no identifiable cause. Finding such reasons is an analyst's job.

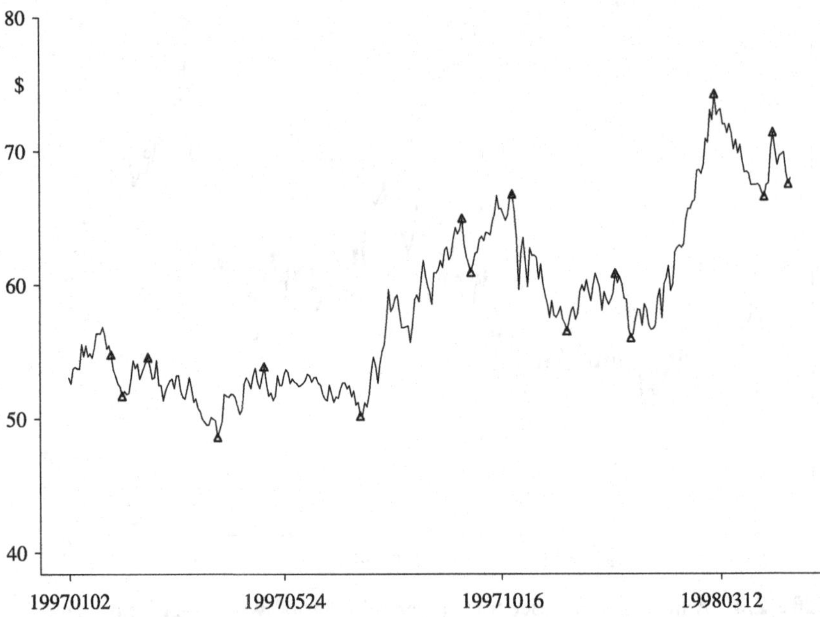

FIGURE 2.7 Adjusted close price trace (General Motors) with 25-percent turning points identified

Table 2.1 gives a summary comparison of the alternative event series for the pair Chrysler (before its acquisition by Daimler) and General Motors. The increase in return correlation for interevent returns is striking, as are the insignificant differences across alternative event series. The latter is a useful property—interevent correlations are robust (not sensitive) to the precise calibration of the event identification algorithm. Therefore, it is not necessary to be overly concerned about which set of events to use in the correlation analysis as a screen for good risk-controlled candidate pairs.

Events in trading volume series provide information sometimes not identified (by turning point analysis) in price series. Volume patterns do not directly affect price spreads but volume spurts are a useful warning that a stock may be subject to unusual trading activity and that price development may therefore not be as characterized in statistical models that have been estimated on average recent historical price series. In historical analysis, flags of unusual activity are extremely important in the evaluation of, for example, simulation

FIGURE 2.8 Adjusted close price trace (General Motors) with 20 percent turning points identified

TABLE 2.1 Event return summary for Chrysler–GM

Criterion	# Events	Return Correlation
daily	332	0.53
30% move	22	0.75
25% move	26	0.73
20% move	33	0.77

results. Identifying volume peaks in historical data is scarcely different from the demonstration of peak identification in price histories documented previously. In live trading, however, forward-looking monitoring for patterns of increased trading volume, an important risk management tool, is subtly different. One needs to flag volume increase during the build-up before a peak is identifiable because identification after the fact is usually too late for ameliorating impact on a portfolio.

2.4.3 Correlation Search in the Twenty-First Century

Several vendors now offer software tools for managing aspects of pairs trading, from identifying tradable pair candidates to execution channels and portfolio management. Correlation searches of the type described here were manually programmed and carried out in the 1980s. No longer is this necessary. Credit Suisse First Boston, for one, offers a tool that allows a user to optimally fit a Bollinger band–type pair trading model to any specified pair of stocks. The program searches over a range of fixed-width windows simulating trading of a mean plus or minus standard deviation model; simulation "profit" is the metric used to compare models (data window length, Bollinger bandwidth) and the maximum profit generating model is identified. One can very quickly fit models to many pairs using such tools. The dangers of relying solely on such shallow data analysis should be immediately evident. Tools with similar capabilities are offered by Goldman Sachs, Reynders Gray, and Lehman Brothers, among others.

At this time, no commercial tools are known to exist that facilitate identification of event or turning points and compute interevent correlations.

2.5 PORTFOLIO CONFIGURATION AND RISK CONTROL

As models were developed, increasing attention was directed to portfolio risk control. Mean–variance approaches were favored for a long time as profits rolled in and risk was deemed "under control." The folly of that thinking was rudely demonstrated in the summer of 1998, but that is getting ahead of the story (see Chapter 8).

Some modelers incorporated risk exposure calculations along with return forecasts into the portfolio construction process directly (see section 2.4, and the description of the defactor model in Chapter 3); others (particularly those whose model comprised a set of rules with no explicit forecast function) first constructed a portfolio, then calculated the exposure of that portfolio to certain defined market factors, and controlled risk by hedging those exposures separately from the bets constituting the portfolio.

The objective is to select a portfolio of stocks that maximizes return to employed capital. Given perfect foresight, the optimal portfolio consists of maximum possible investments in the stocks with the greatest return until available capital is exhausted. Of course, we do not have perfect foresight. In its stead, we make do with the best forecast we have. The goal is still to maximize actual return but, in the forecasting guess world, we have to focus attention on expected return.

Forecasts, unlike foresight, do not come with a guarantee of the outcome. There is risk in acting on forecasts. A single pair spread expected to "revert to its local mean" may continue to increase beyond the point at which stop loss limits force exit from the position. This new element, risk, complicates the goal, which now becomes twofold: Maximize expected return and maintain the risk of achieving that return below a certain tolerance.

So far so good. Going from foresight to forecast we exchange certainty for uncertainty; we move from guaranteed optimization to constrained optimization of a best guess. However, in practice matters are not quite as straightforward as that sentence seems to imply. The first obstacle is precisely specifying the notion of risk—or, at least, its practical implementation. Risk arises because there is no guarantee that a particular forecast will be borne out in reality. Indeed, the truth is that it would be an extraordinary event if a forecast turned out to be 100 percent accurate. Only one outcome yields complete forecast accuracy. But there is an infinity of possible outcomes that translate to odds of infinity to one against the forecast being correct. Hence, the remarkable fact that a forecast is almost certainly going to be wrong.

"Go for the best" becomes "Go for the best guess—but bear in mind what disasters might occur and do your best to protect against those undesirable outcomes."

Just as we have to guess at the best (forecast) we have to guess at the disasters. Typically, we do this a little differently from the way we look for the best guess: Rather than looking for particular disaster scenarios we look at the range of disasters—from small to large—that may befall us. This view is encapsulated in the forecast variance. (Scenario analysis is often used to be aware of "unlikely" extreme situations notwithstanding routine, daily, "risk controlled" portfolio construction. The distinction of *extreme* and *routine* risk is deliberately vague.)

The goal has, therefore, become: Maximize expected return subject to a limit on believed variation about that expected return. The variance constraint reduces the admissible set of portfolios from the set of all portfolios to the set of portfolios for which the expected variation of the expected return is below some threshold.

It is crucial not to lose sight of the fact that all these quantities—forecast returns and variances thereof—are uncertain. The forecast variance guides us as to how much the outcome may reasonably be expected to deviate from the best guess. But that forecast variance is itself a guess. It is not a known quantity. And remember what was stated only two paragraphs ago: Forecast variance characterizes *average* behavior; anything is possible on any given instance.

With all those cautionary remarks having been said, it is true that we are using a forecast constructed in the belief that it has some predictive utility. That, on average—but not in any particular case or set of cases—the forecasts will be better guesses of future events than random guesses. And that the range of variation of outcomes about the forecasts is reasonably quantified by the forecast variances—again, on average.

Finally we are in a position to make operational the notion and quantification of risk. We defined the risk of a portfolio as the expected variance of that portfolio. Our aversion to risk is then taken to be a constant multiple of that variance. Thus, the goal becomes: Maximize expected return subject to a limit on expected variance of return.

Let us express these results in mathematical form. First, definition of terms:

n	Number of stocks in investment universe
f_i	Expected forecast return for stock i; $f = (f_1, \ldots, f_n)'$
Σ	Expected variance of returns, V[f]
i_p	Value to be invested in stock i; $p = (p_1, \ldots, p_n)'$
k	Risk tolerance factor

Now the goal is expressed as:

$$\text{maximize } p'f - kp'\Sigma p$$

2.5.1 Exposure to Market Factors

Statistical arbitrage fund managers typically do not want a portfolio that takes long positions only: Such a portfolio is exposed to the market. (A pairs trading scheme, by definition, will not be biased but statistical arbitrage models more generally readily generate forecasts that, unless constrained, would lead to a portfolio with long or short bias.) If the market crashes, the value of the portfolio crashes with it. This much we can say *regardless* of the precise composition of the portfolio. Given a desire for a market neutral strategy, the goal is to pick off moves in stock prices after allowing for overall market movement. That begs the question of how one defines "the market." Conventionally, the S&P 500 index is taken as (a proxy to) the market. Each stock in the portfolio is statistically examined to quantify the stock's exposure to the S&P index. These quantifications are then used to determine a portfolio's exposure to the market. Market neutrality is achieved by altering the proportions of stocks in the portfolio.

Make the definition:

l_i Exposure of stock i to the market; $l = (l_1, \ldots, l_n)'$

Then the market exposure of the portfolio p is:

$$\text{market exposure} = p'l$$

With the desire for market neutrality, the objective function is modified to:

$$p'f - kp'\Sigma p - \lambda p'l$$

where λ is a Lagrange multiplier (relevant only to the optimization).

The neutrality desire is extended from the market as a whole to include market sectors. We want to avoid overall exposure to, for example, the oil industry. This is accomplished in the same way as is market neutrality: Define exposures of stocks to "the oil industry." Notice that this is a more general notion than simply defining an index for the oil industry and exposures of oil industry stocks to that index. Potentially *every* stock, oil industry or not, has an exposure to the oil industry market factor. Given this set of exposures, the objective function extends in a similar way as for the market factor.

Make the definition:

$l_{1,i}$ Exposure of stock i to the oil industry; $l_1 = (l_{1,1}, \ldots, l_{1,n})'$

The objective function is extended to:

$$p'f - kp'\Sigma p - \lambda p'l - \lambda_1 p'l_1$$

where λ_1 is another Lagrange multiplier.

Obviously, other market factors may be included in the objective function to ensure zero portfolio exposure thereto. For q market factors, the objective function is:

$$p'f - kp'\Sigma p - \lambda p'l - \lambda_1 p'l_1 - \cdots - \lambda_q p'l_q$$

Determining the portfolio that maximizes the objective function is a straightforward application of the Lagrange multiplier method.

2.5.2 Market Impact

We forecast IBM stock to yield annualized return of 10 percent over the next week. The forecast is more certain than any forecast we have ever made. We want to buy \$10 million worth of stock. Ordinarily, a demand of that size will not be filled at the current offer price; most likely the offer price will rise as the demand is filled. This is market impact. Market impact is incurred with most trades, regardless of size, since the market is not static between the time a forecast is made (using the latest available price) and the time the desired trade is placed and subsequently filled. Without actual trading history, it is impossible to gauge market impact. Even with trading history, it is possible only to make a guess: Once again we are forecasting an uncertain event. (See Chapter 10 for recent developments with critical implications for statistical arbitrage.)

The importance of market impact is great. A good estimate of the likely realizable fill price for desired trades enables the trading system to filter potentially unprofitable trades from the portfolio optimization.

Immediately, a question arises: Is market impact not subsumed in the construction of the forecast function? Superficially only. There is an implicit assumption that the stocks can be traded instantaneously at the current price. Okay, but why should the time delay to complete a physical trade result in a cost? Should we not expect that some prices will go with the desired trade and some against, with everything averaging out over numerous trades on many days? Again, superficially only. Our participation in the market is *not* accounted for in the model building process. A buy order from us adds to demand, dragging up price; the opposite for a sell order. Thus, our own trading introduces a force against us in the market. So our forecasts are really only valid providing we do not act on them and participate in the market.

One might ask that, since the goal is to build a forecast model that is exploitable, why not include the information that the forecasts will be traded into the model building? The short—and probably also the long—answer to that is that it is just too difficult. (Equivalently, the necessary data is unavailable; see Chapter 10 for what is possible, even *routine* for a select few, at present.) The pragmatic expedient is therefore to build a forecast that is expected to be valid if we remain a passive observer, then make an adjustment for the effect our active participation is likely to have.

Market impact is a function of what we decide to trade. Denoting the current portfolio by c, the objective function is extended generically to:

$$p'f - \text{market impact}(p - c) - kp'\Sigma p - \lambda p'l - \lambda_1 p'l_1 - \cdots - \lambda_q p'l_q$$

Determining the functional form of "market impact" is an unsolved research problem for most participants because of inadequate data (typically restricted to one's own order and fill records) and, in some cases, lack of motivation. Again, see Chapter 10 for more recent developments.

2.5.3 Risk Control Using Event Correlations

In the preceding section we explored the idea of event correlations as a basis for identifying collections of stocks that existentially share common risk factors: Stocks repeatedly exhibit directional price change at the same time, in the same orientation, and move by a similar amount between such changes. Within a group, stocks have similar elasticity to news or what one might call "event betas."

Building a portfolio that is matched dollar-for-dollar long and short from dollar-matched portfolios built from groups of stocks characterized by similar event betas *automatically* incorporates substantial risk control. Each group defines a collection of stocks that have repeatedly exhibited essentially the same price reaction to economic developments meaningful for those stocks. The key feature here is the *repeated* nature of the moves. To bowdlerize Ian Fleming,[2] *once is happenstance, twice is coincidence, the third time is common risk exposure.* A portfolio thus formed has a low probability of experiencing large loss generating disparate moves of constituent stocks in response to a market shock. After the terrorist attacks on the United States in 2001, event beta–neutral portfolios of large capitalization stocks exhibited only *mundane* volatility in valuation despite the dramatic market decline and spike in individual stock volatility.

2.6 DYNAMICS AND CALIBRATION

The reversion exploitation model is applied to *local* data. For example, estimated interstock volatilities are calculated using a weighting scheme, discounting older data (see Chapter 3). In the trade rules examined earlier in this chapter, we chose a 60-day fixed-length window and computed trade limits on spreads as a function of the spread range (a) directly, and (b) using the empirical standard deviation of the spread distribution. These daily updated estimates adjust current trade entry and exit points. Similarly, daily updated liquidity estimates modify trade deal size and portfolio concentration. Thus, even with an unchanged model there is continuous adaptation to local market conditions.

Occasionally the model is recalibrated (or a manager "blows up"). Recall the CAL–AMR spread, which changed radically from $20 in 2000 to $6 in 2002.

[2]Spoken by Auric Goldfinger to James Bond in Ian Fleming's *Goldfinger*.

The techniques of evolutionary operation (EVOP) can be employed to help uncover persistent changes in the nature of the reversion phenomenon exploited by a model. Reversion is exhibited by stock price spreads on many frequencies (Mandelbrot, fractal analysis), one of which is targeted by a modeler's chosen calibration, that choice being dictated by factors including robustness of the response to small changes in parameter values, modeler's preference, research results, and luck. Applying EVOP, several other frequencies (model calibrations) are monitored in tandem with the traded model to provide information on changes in the nature of the response across frequencies. There is always noise—one frequency never dominates nearby frequencies in terms of actual and simulated trading performance month after month. It is crucial to understand the normal extent of this noise so that apparent (recent) underperformance by the traded model vis-à-vis a nearby (in model space) competitor model is not misinterpreted as a need for a model change. There is also evolution. Over several years, trends in the reversion response as revealed through comparative model performance stand out from the local variation (noise). When identified, such a trend should be adapted to—the traded model calibration revised.

Analysis of a classic pair-trading strategy employing a first-order, dynamic linear model (see Chapter 3) and exhibiting a holding period of about two weeks applied to large capital equities shows a fascinating and revealing development. In March 2000 a trend to a lower frequency that began in 1996 was discovered. First hinted at in 1996, the scale of the change was within experienced local variation bounds, so the hint was only identifiable later. Movement in 1997 was marginal. In 1998, the problems with international credit defaults and the Long Term Capital Management debacle totally disrupted all patterns of performance making inference difficult and hazardous. Although the hint was detectable, the observation was considered unreliable. By early 2000, the hint, there for the fourth consecutive year and now cumulatively strong enough to outweigh expected noise variation, was considered a signal. Structural parameters of the "traded" model were recalibrated for the first time in five years, a move expected to improve return for the next few years by two or three points over what it would otherwise have been. Simulation for 2000–2002 significantly exceeded that expectation as market developments caused a decline in performance of higher frequency compared with lower frequency strategies. See Chapter 9 for a detailed discussion of the issues involved.

2.6.1 Evolutionary Operation: Single Parameter Illustration

Evolutionary operation for a single parameter is illustrated in the four panels of Figure 2.9. Panel (a) shows an archetypal response curve: For a range of possible values for a model coefficient, the (simulated) return from operating the strategy shows a steady increase tailing off into a plateau then quickly falling off a cliff. One would like to identify the value of the parameter for which return is maximized—and that is simple when analyzing past data and when the response relationship is invariant.

Panel (b) illustrates what one observes in practice. Every year the response is different. Similar—that is why strategies work more often than not—but different. When selecting a parameter value at which to operate a strategy, it is critical to understand both the form of the response curve and the natural amount of variation and relate these to understanding of the phenomenon under study—reversion in this case—when it is available. Picking the return-maximizing value of the parameter from panel (a) is risky because in some years the response curve shifts sufficiently that model performance falls off the cliff. Risk management operates at the model calibration stage, too: Back away from the cliff and settle for generally good years and low risk of a catastrophe rather than occasional outstanding years and occasional disasters. One should expect that disasters will occur from uncontrollable factors: Admitting large probabilities of disaster from "controllable" factors is not a sound risk management policy.

Panel (c) shows an archetypal evolution in response: The general form moves smoothly through space over time (and the form may itself smoothly change over time). In practice such evolution, when it occurs, occurs in conjunction with normal system variation as just seen in panel (b). Experience is thus like a combination of the movements in panels (b) and (c), illustrated in panel (d).

As the response curve changes over time, a range of the parameter space consistently yields good strategy performance. Every year is different and over time the parameter range to aim for gradually moves. The original range continues to deliver reasonable performance, but becomes less attractive over several years. Evolutionary

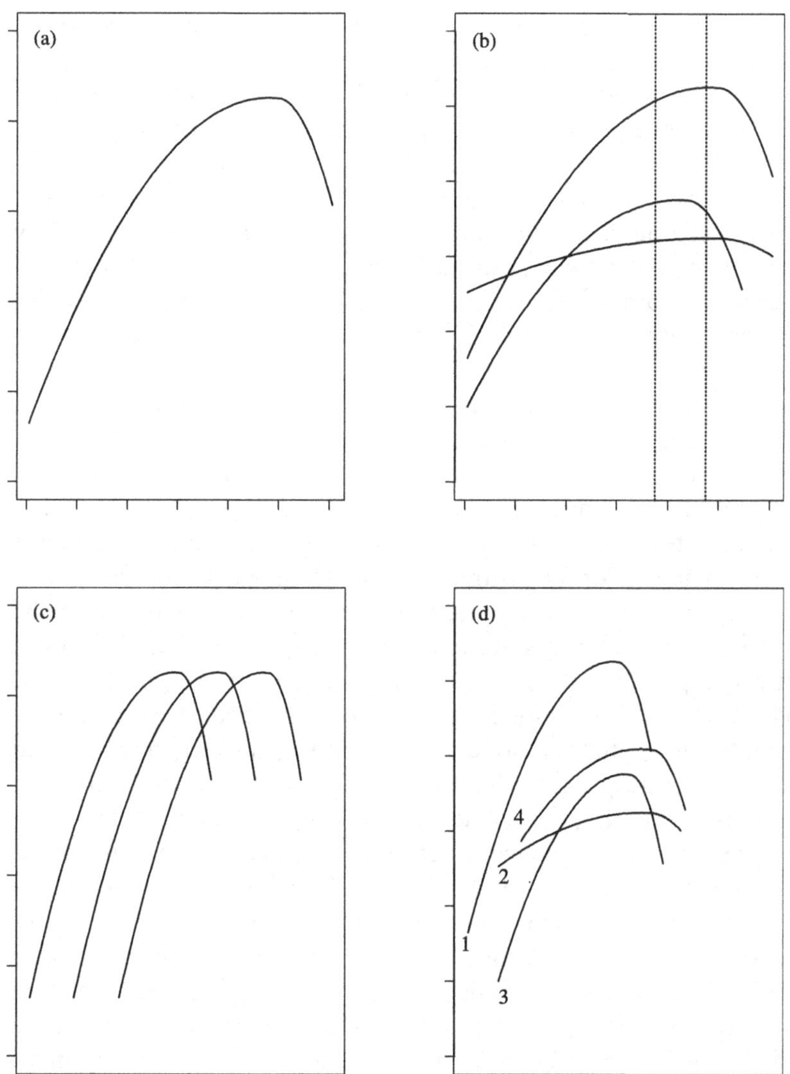

FIGURE 2.9 Evolutionary operation: detecting sustained system response change

operation, the continual monitoring of system performance away from the currently believed best calibration, enables one to identify transient and persistent system response changes. Transient changes provide information to update the view of normal system response variation; enduring system response changes can be adapted to, improving long-term system performance.

As just exemplified, evolutionary operation monitoring looks beguilingly simple—and, indeed, the concepts and rote application are straightforward. Unsurprisingly, reality is more complex. The time scale and timing of change may differ from the annual focus explicitly used here. And, of course, models are typically defined by a collection of parameters, not just one.

Years are somewhat artificial and arbitrary time periods in this monitoring context. Change can occur abruptly within a calendar year (September 11, 2001) or slowly over one or more years. Monitoring aspects of a strategy that reveal diagnostic information at different frequencies is another critical task.

Statistical arbitrage models have several critical defining parameters. The monitoring scheme is complicated because there are interaction effects: The impact of a change in one parameter depends upon the settings of other parameters. The scheme of continual assessment of the performance of alternative model calibrations must be designed to reveal changes in those interactions as well as changes in the direct response of strategy performance to individual parameters.

More complicated models that involve several steps of analysis formally may include hundreds or even thousands of estimated parameters. Conceptually, the monitoring problem is ostensibly the same: One is looking for evidence of change other than transient noise over time. The practice is yet more complicated than for models with handfuls of parameters because these high-parameter-count models often lack manageable interpretability of individual parameters. Answering the question, "What does change X in parameter θ signify?", is impossible. Indeed, with such models even positing the question is difficult. Groups of parameters may have a collective interpretability in which case understanding can be built component by component, sometimes with a hierarchical structure.

To close this section, it is worth reiterating an important point: Underpinning monitoring activity, from mechanics to interpretation to action, is understanding of the phenomenon being exploited—why it exists, what drives the opportunities, and how exploitation works in the context of the model.

CHAPTER 3

Structural Models

Private information is practically the source of every large modern fortune.

—*An Ideal Husband*, Oscar Wilde

3.1 INTRODUCTION

The discussion in Chapter 2 is couched largely in terms of trading rules based on estimates of spread ranges calculated on moving windows of data history. Figure 3.1 shows the bands calculated as mean plus or minus one standard deviation using a window of 60 days for the CAL–AMR spread. (Compare this with Figure 2.4, wherein the limits are calculated using the maximum −20 percent range and minimum +20 percent range over a 60-day window, and review the discussion in Section 2.2.) Implicit in these trading rules is a forecast that the spread will in the near future return to the local mean. Figure 3.2 shows the CAL–AMR spread again, this time with the implied forecast function.

In formal terms the point forecast, or expected value, for every future time period is the currently estimated mean value. Now it is not really believed that the spread will actually equal the mean each time period, or even one time period in the near future. (Obviously the trading rules anticipate systematic variation above and below the mean.) It is simply that the best guess based on using the moving average model is that in the near future, the spread will likely exhibit values centered on the mean. How near is "near" remains, like so much else, unspecified.

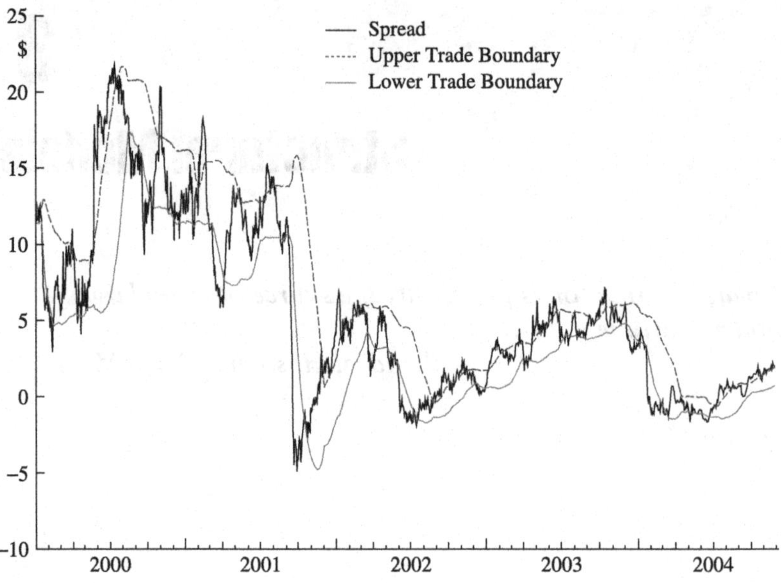

FIGURE 3.1 Daily closing spread, CAL–AMR, with standard deviation trade boundaries

The mean plus or minus standard deviation trading rules were not compiled by formal statistical model building. Rather, simple eyeballing and a little thinking lead to the hypotheses, expressed as trade rules, which turned out to work satisfactorily. Nonetheless, the rules constitute a model with the forecast function interpretation just cited.

The CSFB tool, mentioned in Chapter 2, goes a little further than our eyeballing and systematically searches through many alternative model specifications—window length and number of standard deviations for trade entry boundaries. This model fitting or selection procedure implicitly uses a utility maximization criterion, maximize simulated trading profit, instead of a statistical estimation procedure such as maximum likelihood or least squares. That is a sophisticated approach, unfortunately undermined by the sole focus on in-sample calculations. Effectively, the utility function is misstated for the purpose of identifying models that might be expected to do somewhat reasonably in practice. What is really of interest is maximizing profit out of sample with some regard to draw-down limits, mimicking

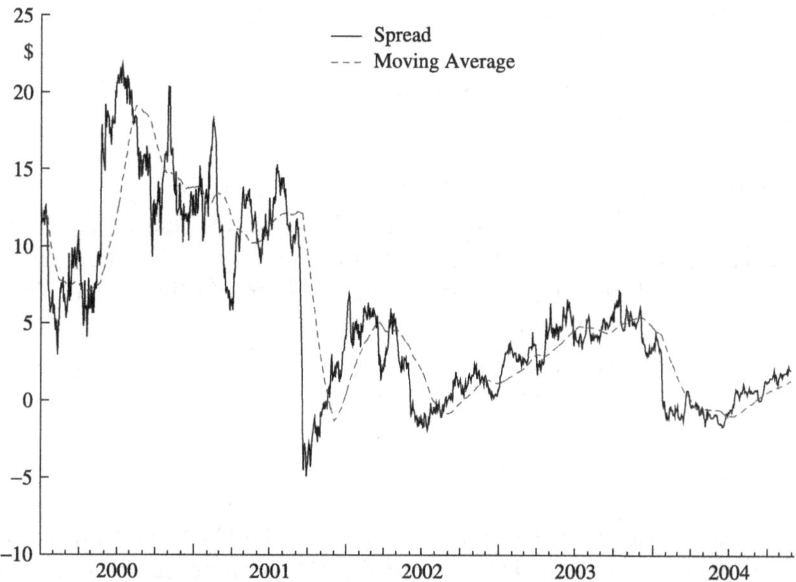

FIGURE 3.2 Daily closing spread, CAL–AMR, with moving average forecast function

actual use of divined trading rules, but those considerations begin to take the tool in the direction of a strategy simulator, which is not likely to be offered free of charge.

3.2 FORMAL FORECAST FUNCTIONS

The value of thinking about a formal forecast function is that it gives a specific set of values to compare to realizations and thereby to judge the efficacy of model projections. Mark to market losses on a trade will indicate the presence of a potential problem; the pattern of forecast–outcome discrepancies provides information on the possible nature of the problem. Such information admits a richer set of responses to loss situations than a blunt stop loss rule such as a simple percentage loss.

In this chapter, we will consider a few of the structurally simplest classical models for time series data. One or two non–time-series-model architectures will also be described illustrating somewhat more involved modeling of stock price data.

3.3 EXPONENTIALLY WEIGHTED MOVING AVERAGE

Moving average, or MA, models are familiar from Chapter 2. A more flexible scheme for local smoothing of series is the exponentially weighted moving average, or EWMA. In contrast to the fixed window of data with equal weights of the MA, the EWMA scheme applies exponentially declining weights to the entire data history. Recent data thereby have the most influence on the current estimate and forecasts, while large events in the remote past retain influence. The form of the projected forecast function (for $k = 1, 2, 3, \ldots, n$ steps ahead) is, like that of the MA, a straight line. The value is different, however. An EWMA is computed recursively as:

$$x_t = \lambda x_{t-1} + (1 - \lambda) y_{t-1}$$

where y_t is the observation at time t, x_t is the EWMA estimate, and λ is the *discount factor*. The size of the discount factor $0 \leq \lambda \leq 1$ dictates how fast older observations become irrelevant to the current estimate (equivalently, how much data history contributes to the current estimate).

The recursive form of the EWMA forecast calculation immediately reveals a simplification over MA schemes. Only the current forecast x_t needs to be retained to combine with the next observation for updating the forecast. While computers don't care whether one or twenty or fifty pieces of information have to be retained in memory, people do. In fact, the moving average can be expressed in a recursive fashion that requires only two pieces of information to be carried so the efficient memory support is unfairly hijacked by EWMA. Much more compelling are the advantages demonstrated below; once familiar with exponential smoothing for forecasting, you will want to consign your moving average routines to the "obsolete" folder.

Figure 3.3 shows the CAL–AMR spread with EWMA(0.04) and MA(60) forecast functions. The EWMA discount factor, 0.04, was selected specifically (by eye—a formal closeness criterion such as minimum mean square could have been employed but this context simply doesn't require that degree of formalism) to give a close match to the 60-day moving average. Only when the raw series (the spread), changes dramatically, do the two forecast functions differ by an

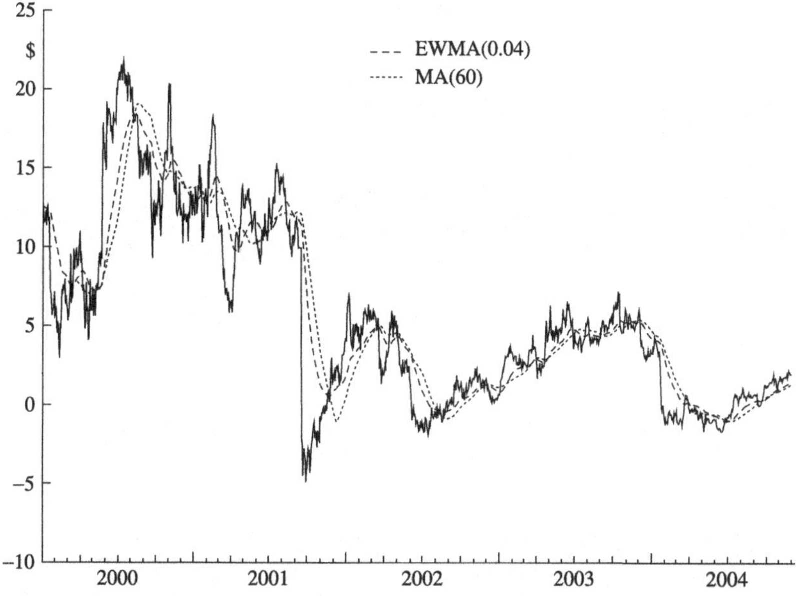

FIGURE 3.3 CAL–AMR spread with EWMA and MA forecast functions

appreciable amount. Table 3.1 gives EWMA discount factors that produce similar local mean estimates to a range of moving averages (for "well behaved" data series).

The utility of the EWMA's flexibility is starkly apparent in two situations where reversion plays fail: step changes and trends in spreads. Figure 3.4 illustrates the situation where a spread suddenly narrows and subsequently varies around the new, lower mean value. The mean and standard deviation bands (using a 20-day window) indicate that the long bet entered on September 7 incurred a large

TABLE 3.1 EWMA–MA equivalences

MA(k)	EWMA(λ)
10	0.20
30	0.09
60	0.04

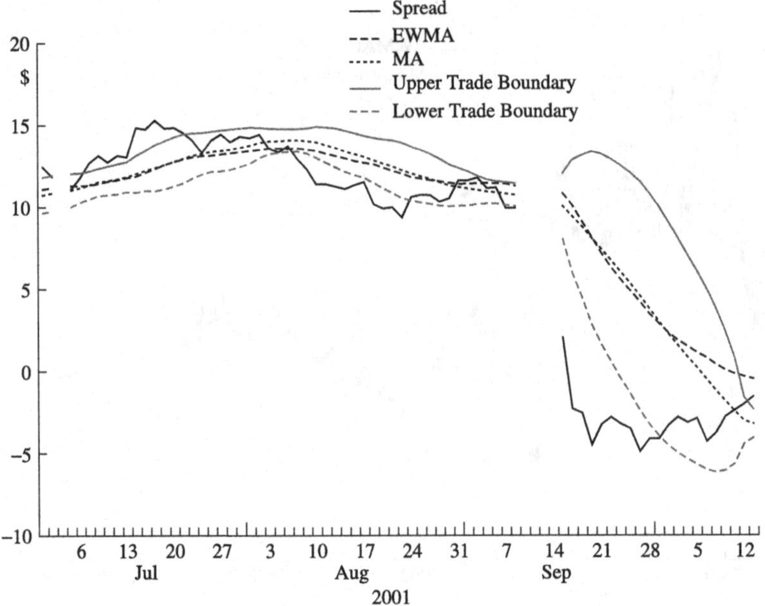

FIGURE 3.4 Level change in spread and MA-EWMA forecast functions

mark to market loss, \$11, at the time of the spread decrease and eventually the bet closed at a loss of that magnitude on October 11 (assuming a popcorn process model). Using the EWMA instead of the MA is of negligible difference. The flexibility advantage shows up as soon as we introduce forecast monitoring and intervention.

When the large forecast error occurs (the day of the unusual spread decrease) the monitoring system is triggered, alerting the modeler to a potential violation of model assumptions, and hence, invalidating the forecasts. Upon further investigation, the modeler might discover a fundamental reason for the spread behavior, which might lead to a decision to terminate the trade. (No search was necessary on September 17, 2001 but decisions on whether to hold or exit bets were critical to manager performance.) Figure 3.5 illustrates a forecast function in which the historical development of the spread is discarded in favor of a single new observation. Forecast uncertainty would typically be large, illustrated by the wide limits.

If no information is discovered, a reasonable action is to watch closely how the spread develops over the next few days (still looking

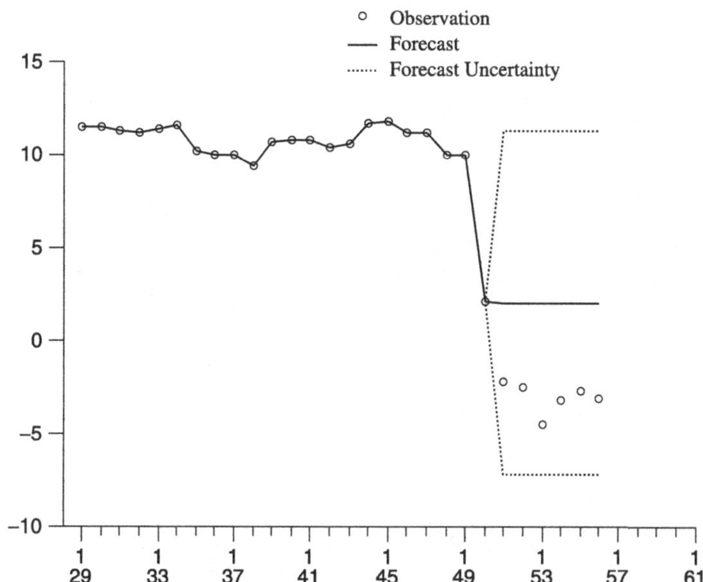

FIGURE 3.5 Intervention forecast function

for fundamental news, of course). If the spread begins migrating back to the pre-shift range, then no action is necessary. If the spread continues to move around the newly established level, then the model forecasts can be improved by introducing that knowledge to the model. With the EWMA it is straightforward to make the adjustment. By increasing the discount factor for just one period, giving more weight to recent spread values, the forecasts quickly become centered around the newly established level, as shown in Figure 3.6. Judgments of the value of the open bet, and of new bets, are improved much sooner than otherwise. The open bet is exited sooner, on September 21, still at a loss but the capital is freed up and the position risk eliminated. Profitable new bets are quickly identified which, without the adjustment, would have been missed while the routine model played catch up: September 27 through October 2 and October 8 through October 10 for a combined gain of $3.64.

Forecast monitoring and model adjustment are also feasible with the MA model but the practicalities of the adjustment are considerably more awkward than the one-time use of an intervention discount factor in the EWMA. Try it and see!

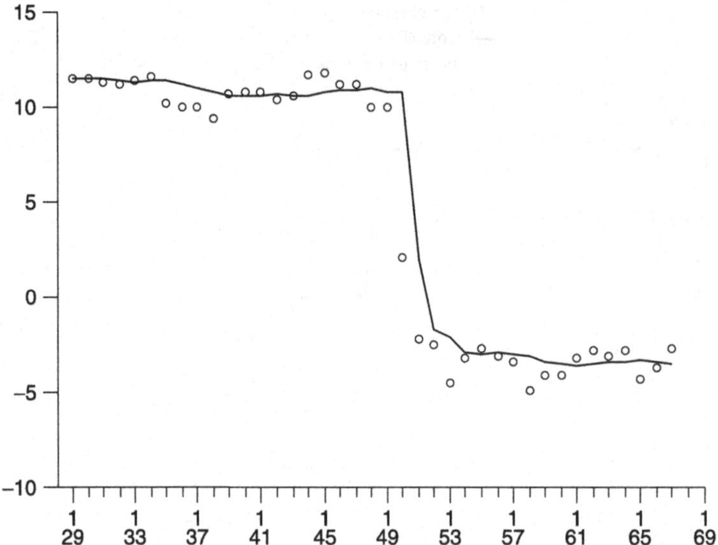

FIGURE 3.6 Post-intervention forecasts

From where did the value of the intervention discount come? Many sophisticated and fascinating control schemes exist in the engineering literature, but for our purposes a simple calibration procedure can be used. From a collection of spread histories, isolate the points at which step changes in the spread occurred. Experiment with a range of intervention discount factors until the pattern of forecasts across all the cases is adequate. (Once more, subjective terms such as "adequate" are left to your interpretation.)

How large a spread is indicative of possible level shift? Look again at your data: Three standard deviations from the mean occurs how often? How many false monitor alarms would ensue with that calibration? What about four standard deviations? How many level shifts are missed? With what consequences for spread bets? It is not useful to rely on probabilities of standard deviation moves for a normal distribution—3 standard deviations from the mean occurring 0.2 percent of the time—because spreads are typically not normally distributed. To see this, form the daily spread data from your favorite pair into a histogram, overlay the best fitting normal density curve (match the sample mean and variance). Examine the quality of the fit in the tails and in the center of the density.

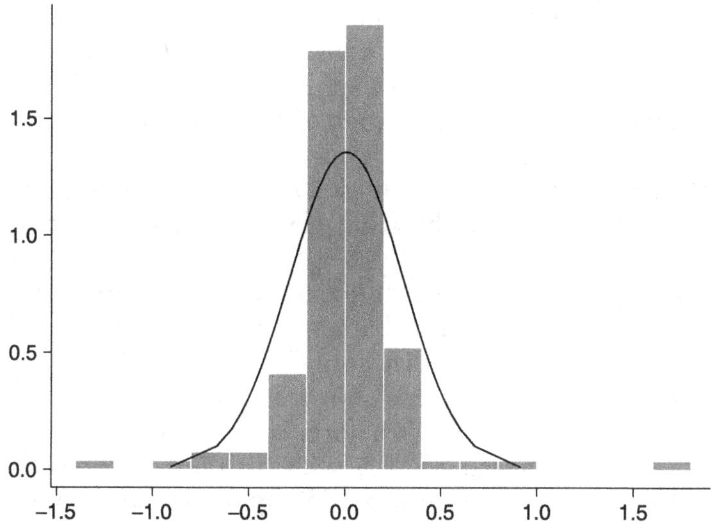

FIGURE 3.7 CAL–AMR spread returns, December 2001 to May 2002, with normal density

Figure 3.7 illustrates a common situation. Daily returns from a sample of six months of the CAL–AMR spread (December 2001 through May 2002) are displayed in a histogram; a normal density curve is fitted to the sample mean and standard deviation is superimposed. I leave the commentary to you.

Empirical experimentation is a sounder approach to developing understanding than blind assumption of normality, and a good place from which to build a representative formal model if that is your goal. Chapter 5 debunks some common misconceptions about underlying distributions and reversion in time series data.

A more detailed discussion of forecast monitoring; intervention and automatic adaptation schemes, including likelihood based tests instead of the rudimentary standard deviation rules suggested here; and evidence accumulation strategies is given in Pole et al., 1994. That volume details the class of models known as dynamic linear models (DLM) which contain as special cases MA and EWMA models, and also autoregressive models that feature prominently in some statistical arbitrageurs' offerings. The structure of the DLM provides for a rather richer analysis than anything discussed in this volume.

In Chapter 9, Section 2, a situation that practically screamed for intervention is related: the forced liquidation of \$4.4 billion in October 2003 because of massive redemptions in response to the New York attorney general's investigation of Janus for mutual fund market timing activities. Expected market reaction to the terrorist attacks on the United States in September 2001 is a fine example of the need for careful review and the value of well designed, *selective* intervention.

Not all changes of level are as dramatic as in the preceding example. Often a new level is reached after a migration over several days rather than in a single, outsize leap. The British Petroleum (BP)–Royal Dutch Shell (RD) spread shown in Figure 3.8 exhibits several such migrations. Two EWMA forecast functions are illustrated. The first is a standard EWMA with discount factor 0.09 (which is similar to a moving average on 25 days except at times of significant change where the MA lags the EWMA's adjustment to the data, as previously remarked), which adapts rather slowly to the

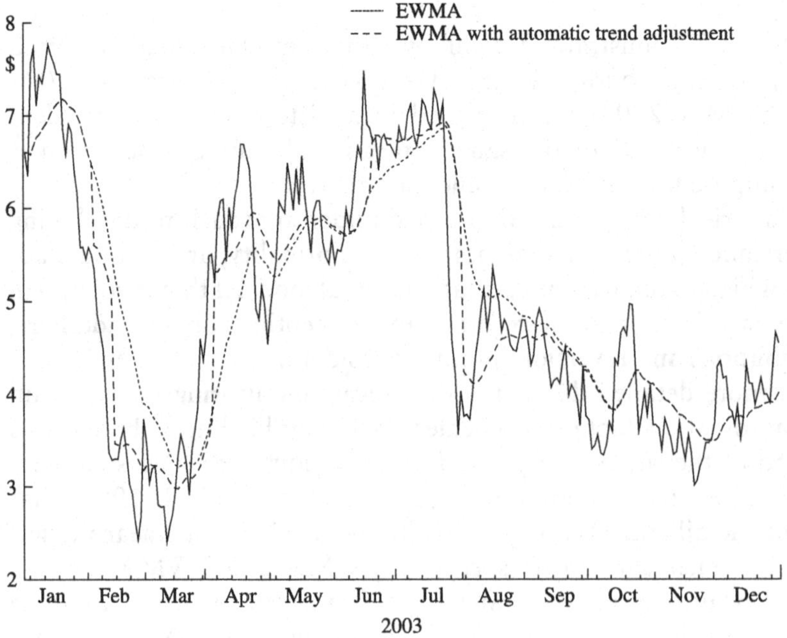

FIGURE 3.8 Trend detection and level adjustment

\$4 change in spread level in the first quarter of 2003. The second is an EWMA that switches to a high discount factor when a level shift is detected to be underway. The increased pace of adaptation is evident in the downward shift of the spread in February 2003, the two subsequent shifts upward in April and June, and another downward shift in late July.

For this demonstration, the switching rule is quite crude: When the spread exceeds a one standard deviation margin from the basic EWMA for several days in succession, use the high discount factor for faster adjustment. (The local standard deviation is computed using EWMA smoothing.)

Before leaving the BP–RD spread, take another look. Over the whole year, the spread exhibits a nicely controlled sinusoidal-like variation about a mean of \$5. What do you make of that?

3.4 CLASSICAL TIME SERIES MODELS

There are many books on the market that describe models for characterizing time series and making forecasts. A few are listed in the bibliography and those are your first stopping point for a detailed understanding of model forms, statistical estimation and forecasting procedures, and practical guidance on data analysis and model building. In this section we give heuristic descriptions of several model types that have been successfully employed by statistical arbitrageurs. The discussion will be grounded in the context of the spread and reversion descriptions and archetypal underlying processes (sinusoidal and popcorn).

3.4.1 Autoregression and Cointegration

Probably the most commonly applied time series model structure in any field is the autoregressive model. Future values of a series are projected as weighted averages of recently exhibited values in that series. Examples already seen include the moving average and the exponentially weighted moving average.

The autoregressive model of order p relates the series outcome at time t to a linear combination of the p immediately preceding outcomes:

$$y_t = \beta_1 y_{t-1} + \cdots + \beta_{t-p} y_{t-p} + \epsilon_t$$

The coefficients, β_i, or model parameters, are determined by estimation from a set of observations of the series. The final term, ϵ_t is the so-called "error term" in which is conveniently gathered all the variability that cannot readily be subsumed in the structural part of the model. This is the term typically assumed to be normally distributed when standard estimation methods are applied and properties of parameter estimates or forecasts are discussed.

Autoregressive models often appear as ARIMA models, which stands for autoregressive integrated moving average. The classic, and unsurpassed, reference is Box and Jenkins (1976). A moving average model in this context is, confusingly, somewhat different from the moving average previously encountered. Here it is really a shorthand way of writing a very long moving average of a series past values using a mathematical equivalence to an average of a few imaginary terms. That is quite a mouthful as well as brain fag so we will not pursue it here.

And what of the "integrated" part? That is simply a differencing operation applied to a series before investigating autoregression structure. For example, daily differences in prices, $z_t = p_t - p_{t-1}$ in an obvious notation, might exhibit autoregression structure. The model for the raw price series is then called an integrated autoregression. The EWMA forecast function, while originally developed in the logical, data analytic way for smoothing variable observations as we introduced it earlier, is actually derivable as the optimal forecast for an integrated model.

This leads nicely to cointegration. Often several series are observed to move together in ways suggestive of a relationship; common situations include (a) one series driving another, and (b) several series driven by common underlying processes. Multivariate forms of ARIMA models can represent very complicated structures of this sort including contemporaneous and lagged feedback relationships.

A structure familiar to spread modelers (but perhaps not known by its technical name) is cointegration. Where two (or more) series are nonstationary individually but their difference (the spread, in our context) is stationary (think of that as meaning "approximated in good measure by an autoregression"), the series are called cointegrated. The difference (and it may be a difference other than the first though we will not pursue that here) is well modeled by an autoregression.

A related class of autoregression models provides parsimonious structural forms for series with long-term serial correlation dependencies. Long-term correlation can be directly captured by a very high order autoregression, but estimation problems ensue because of the high parameter count. Autoregressive fractionally integrated moving average (ARFIMA) models overcome the parameter count problem, essentially fitting ARMA models to series after fractionally differencing.

3.4.2 Dynamic Linear Model

All of the models discussed in the preceding section rely on a considerable degree of stationarity in data series for their efficacy. Model parameters are estimated over a long history of data and are supposed to be unchanging. In financial practice, it is seldom that relationships in data are even approximately unchanging for any length of time. Parameter updating procedures are universal, refitting models to moving windows of data, a commonly used (and useful) device. The local mean and volatility calculations used in Chapter 2 exemplify the procedure.

A flexible model structure that directly embodies temporal movement in data defining qualities, local mean for example, is the dynamic linear model. In the DLM, temporal change in model parameters is explicitly included through the specification of an evolution equation. Consider a first order autoregression:

$$y_t = \beta y_{t-1} + \epsilon_t$$

in which the realizations of a series are composed of two parts: a systematic propagation of a fixed portion of the immediate past defined by the parameter β, and a random addition ϵ_t. Now consider a flexible generalization of that model in which the systematic element propagated sequentially may vary in magnitude period by period. The parameter β is now time indexed and its variation is strictly formalized so that evolution is permitted but revolution is not. The dynamic model is specified by two equations, one defining the observation sequence and one defining the systematic evolution:

$$\begin{aligned} y_t &= \beta_t y_{t-1} + \epsilon_t && \textit{observation equation} \\ \beta_t &= \beta_{t-1} + \omega_t && \textit{system equation} \end{aligned}$$

In the system equation, the term ω_t is a random term that controls, by the magnitude of its variance, how fast the regression coefficient β_t can change. With ω_t identically zero, the dynamic model reduces to the familiar static model. With a "large" variance for ω_t, the data series history is immediately discounted so that $\beta_t = y_t/y_{t-1}$. You may begin to see how intervention in the dynamic linear model, exemplified for the EWMA model in Section 3.3, is implemented.

The DLM includes ARIMA, EWMA, and regression models as special cases, making it a rich, flexible class with which to work. Monitoring and intervention strategies are readily defined for each model component separately and in combination. See, Pole, et al. for examples.

3.4.3 Volatility Modeling

Volatility modeling has an extensive pedigree in quantitative finance. Use in statistical arbitrage is less direct than in derivative valuation where most theoretical development and published applications are seen, but it is nonetheless helpful. Consider just the simple spread modeling that provides much of the background of the discussion in this book: The variance of the return stream determines the richness of potential bets (the basic viability of candidate raw material for a strategy), variability of mark to market gains and losses while a bet is extant (the risk profile of a strategy, stop loss rules), and return stretching by stochastic resonance (see Section 3.7).

Generalized autoregressive conditional heteroscedastic (GARCH) and stochastic volatility models loom large in the modeling of volatilities. The derivatives literature is replete with variants of the basic GARCH model with acronyms ranging from AGARCH through EGARCH to GJR GARCH, IGARCH, SGARCH and TGARCH. GARCH models are linear regression models with a nonlinear structural specification for the error variance. The error variance, in other models assumed to be a constant or a known function of some aspect of the model or time (see the discussion of variance laws in Pole, et al.), is specified to be a linear function of the error term in the basic regression function. Consider, again, the first-order autoregression and now add a first-order GARCH component:

$$\begin{aligned} y_t &= \beta y_{t-1} + \epsilon_t, \qquad e_t \sim N(0, h_t), \\ h_t &= \alpha_0 + \alpha_1 \epsilon_{t-1}^2 \end{aligned}$$

The notation, $e_t \sim N(0, h_t)$ means that the (error) term e_t is assumed to be randomly distributed according to the normal distribution with mean 0 and variance h_t. (The "TGARCH" model uses a Student t distribution in place of the normal for smoother response to "large" deviations.) In this model, the disparity between a forecast and the corresponding outcome feeds directly into subsequent forecast variance. A large forecast error "forces" the model to anticipate forthcoming large volatility. That, in turn, means that less weight will be accorded to the next observation in updating parameter estimates. Therefore, when a model with GARCH error structure is fitted to data that exhibits volatility clusters (bursts of higher than normal volatility) the weight given to the more variable observations in estimating the structural part of the model is reduced relative to the weight given to less variable observations.

In contrast to weighted estimation procedures, which assume a known functional form of variance variability (such as the "level to the power 1.5" which arises often in product sales data being the result of a compound Poisson process), the GARCH model estimates the changing pattern of variation along with the structural part of the model. The pattern of variability is not specified ahead of time, but a rule of recognition is: Large forecast–outcome discrepancies signify large volatility.

Models may include greater lag structure—more ϵ_{t-k} terms much like higher order autoregression models for mean structure. Interpretation of such models is difficult and, unsurprisingly, successful applications are largely restricted to low lag structures.

There is an enormous literature on GARCH models, beginning with Engle's 1982 paper, with applications in macroeconomics and finance.

3.4.4 Pattern Finding Techniques

Exploiting persistent patterns of stock price behavior has been approached directly through pattern finding procedures including neural networks and wavelets. Wavelet analysis, a sort of localized Fourier analysis, decomposes a time series into a collection of locally orthogonal basis functions with weights appropriate to the raw series in question. A neural network is a collection of weighted transformation functions; there is no explicit temporal structure but such structure is implicit in the transformation of the inputs (past observations of a series) to the output (forecasts).

Neural networks are excellent tools for finding patterns in data. Where patterns recur, network forecasts can be extraordinarily good. A major drawback though is the lack of interpretability. While it is possible to disentangle the transformations in a small network (at most a single hidden layer, only a handful of nodes per layer, well behaved transfer functions) and thereby attach theoretical understanding, this is not the routine situation. And what of that? If a neural network successfully identifies predictive footprints in stock price data, what does it matter if the intellectual grasp of the input–output transformation is looser than a competitor's model built from autoregressions applied to (say) factor residuals (Section 3.6)? It may not matter at all and we leave the contentious matter to your consideration.

A great advantage of neural networks is their flexibility, which is the reason they are such good pattern identifiers to begin with. When structural change occurs, neural networks can be very quick to identify that a change is underway and subsequently characterize newly stable patterns. The attendant danger, always a partner of such flexibility, is that identified patterns may be ephemeral, their existence fleeting in terms of usable exploitation opportunities. Borrowing from Orwell's fancy: description yes, prediction no.

3.4.5 Fractal Analysis

We refer the interested reader to the inventor, Benoit B. Mandelbrot 2004, who tells it best.

3.5 WHICH RETURN?

Which return do you want to forecast? The answer may seem obvious if you have a particular context in mind: Forecast return for today and trade around that. In general the answer is not obvious outside the context of theory, data analysis, and trading goals. Let's assume that the latter is simply to maximize strategy return (subject to some risk controls that will be left unspecified here). Without theoretical guidance, we might proceed simply to explore some traditional time scales, investigating patterns of daily, weekly, or monthly return.

A little more thought might suggest investigating how return evolves with duration: Examining returns for $1, 2, 3, \ldots, k$ days might indicate a natural time period for a particular type of series, whether it is individual raw stock prices or functions thereof such as factors (see Section 3.6); one might also want to examine the maximum return over the next m days.

Pattern matching models, more elaborate and technically demanding than the models discussed in this book, lead one to consider more general, multivariate functions of stock return series.

3.6 A FACTOR MODEL

The modeling discussion thus far has focused on spreads between pairs of stocks, the domain where statistical arbitrage, as pairs trading, had its genesis. Now we will discuss some modeling ideas applied to individual stock price series analyzed as a collection.

The notion of common risk factors, familiar from Barra-type models, lies at the heart of so-called factor models for stock returns: The basic idea is that returns on a stock can be decomposed into a part that is determined by one or more underlying factors in the market (and in common with other stocks) and a part that is specific to the stock, so-called idiosyncratic return:

$$\textit{stock return} = \textit{return to market factors} + \textit{idiosyncratic return}$$

Early models formed using this decomposition simply identified market factors as industries (the S&P industry sectors) and a general market factor. Some modelers used indexes as proxies for the factors, building multiple regression models, autoregression models, or other models for daily, weekly, or monthly returns; they also fashioned forecasts from (a) forecast models for indexes (b) the constructed regression (etc.) models, and built portfolios accordingly.

Later attempts used a more general model called a statistical factor model. In a factor model, the factors are estimated from the historical stock return data and a stock's return may be dependent on, or driven by, several of these factors.

3.6.1 Factor Analysis

A factor analysis of a multivariate data set seeks to estimate a statistical model in which the data are "explained" by regression on a set of m factors, each factor being itself a linear combination (weighted average) of the observables.

Factor analysis has much in common with principal component analysis (PCA) which, since it is more familiar, is good for comparison. Factor analysis is a model based procedure whereas principal component analysis is not. PCA looks at a set of data and finds those directions in observation space in which the data exhibits greatest variation. Factor analysis seeks to estimate weights for a set of linear combinations of the observables—so-called factors—to minimize discrepancy between observations and model fitted values.

If the distinction between PCA and factor analysis seems hazy, good. It is. And we will say no more about it.

Suppose the universe of stocks has p elements (stocks). We might usefully entertain the component stocks of the S&P 500 index (as of a fixed date) as an orienting example. Pick a number of factors, m. Given daily historical returns on the selected stock universe, a factor analysis procedure will yield m factors defined by their respective *factor loadings*. These loadings are weights applied to each of the stocks. Thus, factor 1 has loadings $l_{1,1}, \ldots, l_{1,500}$. The other $m - 1$ factors similarly have their own loadings.

Multiplying the loadings by the (unobserved) factors yields values for the returns. Thus, given the loadings matrix L, the columns of which are the loadings vectors just described, estimates of the factors, or *factor scores*, can be calculated.

So, after a factor analysis one has, in addition to the original p stock return time series, m time series of factor estimates. It may help to think of the parallel with industry index construction; some statistical factors may look like industry indexes, and may even be thought of that way. But keep in mind the important structural distinction that statistical factors are a function solely of stock price history with no information on company fundamentals considered.

If one were to regress the stock returns on the factors one would obtain a set of regression coefficients. For each stock, there is one coefficient per factor. These coefficients are the stock *exposures* to the factors. By construction of the factors, there is no other set of m linear combinations of the observables that can give a better regression for the chosen estimation criterion (most often maximum likelihood). There are infinitely many *equivalent* sets, however. Strategies with names such as varimax rotation and principal factors—related to principal component analysis—are used to select a unique member from this infinite set.

Note that there is a duality between factor loadings and stock exposures to factors. The duality, which is a consequence of the factor definition and construction, is such that the rows of the loadings matrix are the stock exposures to the factors. That is, in the $p \times m$ loadings matrix L, the element $l_{i,j}$ is both the loading of the jth factor on the ith stock and the exposure of the ith stock to the jth factor.

What is the interpretation of a factor model? It is this: The universe of p stocks is supposed to be a heavily confused view of a much smaller set of fundamental entities—factors. Crudely, one might suppose that the stock universe is really driven by one factor called "the market." Less crudely one might suppose that, in addition to the market, there are a dozen "industry" factors. The factor analysis may then be viewed as a statistical procedure to disentangle the structure—the factors—from the noisy image presented by the full stock universe, and to show how the stock universe we observe is constructed from the "real" factor structure.

3.6.2 Defactored Returns

Another successful model based on factor analysis reversed the usual thinking: Take out the market and sector movements from stock returns *before* building a forecast model. The rationale is this: To the extent that market factors are unpredictable but sentiment about the relative position of individual stocks is stable over several days, such filtered returns should exhibit more predictable structure. Let's look in a little more detail at this interesting idea.

Residuals from the fitted regression model (stocks regressed on estimated factors) will be referred to as defactored returns. It is these defactored returns on which attention is focused. Why? The notion is that return to a stock may be considered as being composed of return to a set of underlying factors (market, industry, or whatever other interpretation might be entertained) plus some individual stock-specific amount. For a stock i this may be expressed algebraically as:

$$r_i = r_{f_1}, + \cdots + r_{f_m} + r_{s_i}$$

For a market and industry neutral portfolio, it is the stock-specific component which remains in the residual of the standard fitted model. Moreover, the stock-specific component may be more predictable than the others in the short term and following this construction. For example, regardless of today's overall market sentiment, the relative positions (value) of a set of related stocks are likely to be the similar to what they were yesterday. In such situations, a portfolio constructed from forecasts of "de-marketed" returns is still likely to yield a positive result.

A simplified illustration may help to convey the essence of the notion. Suppose a market comprises just two stocks in roughly equal proportion (capitalization, value, price). On day t the return may be denoted:

$$r_{1,t} = m_t + \eta_t,$$
$$r_{2,t} = m_t - \eta_t$$

In this case, the factor model will include just one component and historical data analysis will reveal the market to be essentially the average of the two constituent stocks. (More generally, different stocks will have different exposures to a factor—m_t would appear weighted in the equations—but with weights intimately bound up with the factor definition as already described.) In this case, the stock-specific return will be of the same magnitude for each stock, but signed differently. Now, if this quantity, η_t, can be predicted better than by random guess, then regardless of the pattern of the market return, a portfolio long stock 1 and short stock 2 (vice versa when η is negative) will, on average, yield a positive return.

In a more realistic situation, as long as many bets are made and there is some forecast power in the defactored return model (which may be a EWMA, autoregression, etc.), the trading strategy should win. Making bets dependent on the size of forecast returns and optimizing selected portfolios for risk should improve return/risk performance.

Brief details of the algebra of factor analysis and the construction of the defactored returns model are given in Section 3.10.

Operational Construction of Defactored Returns Factor loadings/exposures must be updated periodically to maintain any reasonable expectation of forecast (and hence trading) performance. Since the statistical factors directly reflect (supposed) structure in the stock price histories, it is not surprising to discover that the structure is dynamic. Estimating factor relationships from stale data will most likely produce results with unpromising forecast performance. The selection of the frequency of factor updating is, like similar dynamic model elements previously remarked on, a matter for the investigator's art. Quarterly or half yearly revision cycles are often used.

Defactored returns must be calculated using the most recent *past* set of loading estimates and *not* the contemporaneous set, ensuring that the defactored return series are always defactored out of sample. While this is deleterious for simulation results, it is critical for strategy implementation. It is easy to pay lip service to this commonly acknowledged matter but also easy in a complicated model or estimation procedure to forget it.

A dynamic model, generalization of the DLM might be considered so that model parameters are revised each day according to a structural equation, but the extra computational complexity was not justified in the late 1980s. Today there is no such computational concern and dynamic factor models have appeared in the statistical literature with applications to stock price prediction. With these complicated models it is incredibly easy, and tempting, to allow great flexibility, unwittingly taking a path to a model that does little more than follow the data. More than one manager eventually fell victim to the seduction, optimized to oblivion.

3.6.3 Prediction Model

After all the work necessary to build the time series of defactored returns for each stock, the modeler is still faced with constructing a forecast model for those returns. That does not imply a return to first base since that would mean that the rationale for the defactorization was void. Nonetheless, one is, as stated, faced with a forecast model building task. One might consider autoregressive models, for example. Note that cointegration models should presumably be of little value here because common factors are supposedly removed in the defactorization procedure.

Many elaborations may be entertained. For instance, there may be more stability in factor estimation on time scales with granularity much longer than one day.

The natural alternative of building a forecast model for the factor series and predicting those series may be entertained. However, this would not replace the defactored return predictions: In the simple example of the previous section, factor prediction is equivalent to prediction of m_t (or a cumulative version thereof). The defactored component is still present.

An unanswered question that arises in this consideration of return forecasting is: What is the relationship between k-day ahead cumulative stock returns and k-day ahead factor estimates? From the earlier discussion, another pertinent consideration is: If, as posited, the market factors are more erratic than the defactored component, then the forecasts will be less useful (in the sense that trading them will yield more volatile results). These considerations indicate that factor predictions are a secondary task for return exploitation (in the context of a valid defactor model). However, factor prediction models—defactor structured return model or not—are useful in monitoring for market structural change and identifying the nature and extent of such change.

3.7 STOCHASTIC RESONANCE

With a model-based understanding of spread or stock price temporal dynamics, there is another crucial part of the process in which analysis can demonstrate exploitation possibilities. Consider a spread that may be characterized as a popcorn process: Occasionally the spread departs from its (locally in time) "normal" value subsequently to return to that norm over a reasonably well defined trajectory. The normal level is not constant. When not subject to some kind of motion inducing force such as a block trade, spreads meander around a local average value, sometimes greater and sometimes less. This motion is largely random—it can, at least, be satisfactorily considered random in the present context. Knowing that once a spread has "returned" to its mean it will henceforth exhibit essentially random variation about that mean suggests that the reversion exit rule can be modified from the basic "exit when the forecast is zero" to "exit a little on the other side of the zero forecast from which the trade was entered." Here the "little" is calibrated by analysis of the range of variability of the spread in recent episodes of wandering about the mean before it took off (up or down). Volatility forecasting models, GARCH, stochastic volatility, or other models may be useful in this task.

The phenomenon of "noise at rest," the random wandering about the local mean just exemplified, is known as stochastic resonance.

As you read the foregoing description, you may feel a sense of *deja vu*. The description of modeling the variation about the mean during periods of zero forecast activity is quite the same as the general description of the variation of the spread overall. Such self-similarity occurs throughout nature according to Benoit Mandelbrot, who invented a branch of mathematics called fractals for the study and analysis of such patterns. Mandelbrot, 2004, has argued that fractal analysis provides a better model for understanding the movements of prices of financial instruments than anything currently in the mathematical finance literature. It is unknown whether any successful trading strategies have been built using fractal analysis; Mandelbrot himself does not believe his tools are yet sufficiently developed for prediction of financial series to be feasible.

3.8 PRACTICAL MATTERS

Forecasts of stock price movements are incredibly inaccurate. Take this message to heart, especially if you have completed a standard introductory course on statistical regression analysis. The traditional presentation proclaims that a regression model is not very useful (some statisticians would say useless) if the R-square is less than 70 percent. If you have not taken such a course and do not know what an R-square is, no matter: Read on. The traditional presentation is not wrong. It is just not appropriate to the situation we are concerned with here. Now, observing that your weekly return regressions produced fitted R-squares of 10 percent or less, perk up!

The key to successfully exploiting predictions that are not very accurate is that the direction is forecast correctly somewhat better than 50 percent of the time (assuming that up and down forecasts are equally accurate).[1] If a model makes correct directional forecasts $(50+\epsilon)\%$ of the time, then the net gain is $(50+\epsilon)-(50-\epsilon)\% = 2\epsilon\%$ of the bets. This net gain can be realized if one can make a sufficient number of bets. The latter caveat is crucial because averages are reliable indicators of performance only in the aggregate.

Guaranteeing that $2\epsilon\%$ of one's bets is the net outcome of a strategy is not sufficient, by itself, to guarantee making a profit: Those bets must cover transaction costs. And remember, it is not the

[1]The situation is actually more complicated in a manner that is advantageous to a fund manager. Symmetry on gains and losses makes for a simple presentation of the point that a small bias can drive a successful strategy; one can readily live with relative odds that would cause a physician nightmares. The practical outcome of a collection of bets is determined by the sum of the gains minus the sum of the losses. A big win pays for many small losses. The significance of this fact is in directing a manager to construct stop loss rules (early exit from a bet that is not working according to forecast expectation) that curtail losses without limiting gains. Where this is possible, a model with seemingly textbook sized relative odds in favor of winning forecasts can be profitably traded within prescribed risk tolerances. Technically, such rules modify the utility function of a model by altering the characteristics of the outcome set by employing a procedure in which the forecast model is only one of several elements.

A warning: Beware of being fooled by purveyors of tales of randomness. A strategy that offers bets that typically generate a small loss and occasionally a whopping gain sounds alluring when proffered as relief after a cunningly woven web of disaster shown to seemingly inevitably follow plays where the odds are conventionally in favor of winning. After these examples of catastrophe are depicted, solace is offered in the guise of an alternative characterized by low risk (small losses) with large gain potential. A crafty invocation of the straw man technique of persuasion. Or, statisticulation, as Huff would call it.

A fine artisan of the word weaves an impressing story of unavoidable doom employing unimpeachable calculus of probability. Then, Pow! Batman saves the "What can I do?" day with a tale of the occasional big win bought by easy-to-take small losses. A complete reversal of pattern. That cannot—can it?—but dispel the doom instantly. Opposite pattern must beget opposite emotion. Joy!

Now about those small losses. Lots of small losses. Total up those small losses and discover the shamelessly omitted (oops, I mean inadvertently hidden in the detail) large cumulative loss over an extended period before the Batman surprise. So what have we truly gotten? A few periods of glee before inevitable catastrophe supplanted with prolonged, ulcer inducing negativity, despondency, despair, and (if you can stand the wait) *possible* vindication! It is still an uncertain game. Just different rules.

There are many kinds of randomness.

average transaction cost that must be covered by the net gain. It is the much larger total cost of all bets divided by the small percentage of net gain bets that must be covered. For example, if my model wins 51 percent of the time, then the net gain is $51 - 49 = 2$ percent of bets. Thus, out of 100 bets (on average) 51 will be winners and 49 will be losers. I make net 2 winning bets for each 100 placed. Statistically guaranteed. My fee for playing, though, is the fee for making all 100 bets, not just the net 2. Thus, my 2 percent guaranteed net winners must cover the costs for all 100 percent of the bets.

Statistical forecast models can do much more than simply predict direction. They can predict magnitude also. Consider a first-order autoregressive model for weekly returns, for example: The size of the return for next week is forecast explicitly (as a fraction of the return for last week). If an estimated statistical model has any validity, then those magnitudes can be used to improve trade selection: Forecasts that are smaller than trade cost are ignored. No point in making bets that have expected gain less than the cost of the game, is there?

Now, what about all that prediction inaccuracy we talked about? If predictions are okay on average but lousy individually, how can we rely on individual forecasts to weed out trades with expected return lower than trade cost? Won't we throw away trades that turn out to be enormously profitable? And take trades that return less than costs?

Indeed yes. Once again, it is the frequency argument that is pertinent here. On average, the set of trades discarded as unprofitable after costs has expected return lower than trade cost. Also, on average, the set of retained trades has expected return greater than trade cost. Thus, the statistically guaranteed net gain trades have expected return greater than trade cost.[2]

3.9 DOUBLING: A DEEPER PERSPECTIVE

It is tempting after an extended discussion of technical models, even at the limited descriptive level of this chapter, to be seduced into

[2]Recall footnote 1, on improving the outcome of a forecast model by imposing a bet rationing (stop loss) rule. Such a procedure increases the average gain of bets made according to the forecast model, so one might squeeze just a little more from an opportunity set by realizing that return bias can convert some raw losing trades (those with average gain less than transaction cost) into winning trades. Subtle. And nice. See also the discussion of stochastic resonance in Section 3.7.

forgetting that models are wrong. Some are useful and that is the context in which we use them. When applying a model, an activity of signal urgency and import is error analysis. Where and how a model fails informs on weaknesses that could be ameliorated and improvements that might be discovered.

In Chapter 2 we introduced "Rule 3," a bet doubling scheme in which a spread bet was doubled if the spread became sufficiently large. The idea was motivated by observing spread patterns in the context of an already formulated model, Rule 1 or 2—this is error analysis notwithstanding the quasi informality of rule positing based on eyeballing data rather than formal statistical model building.

With more complicated models, eyeball analysis is infeasible. Then one must explicitly focus upon the results of trading a model, either in practice (with dollars at risk) or synthetically, (using simulations). Beyond the standard fare of comparing forecast return with outcome one can examine the trajectory of bet outcome from the point of placement to the point of unwinding. In the example of the spread doubling the typical trajectory of cumulative return on the original bet is a J curve: Losses at first are subsequently recovered then (the doubling phase) profits accrue. Trade analysis from any model, regardless of complexity, can reveal such evolutionary patterns and, hence, provide raw material for strategy enhancements such as doubling.

Notice how the dynamic of the trade, not the identification of placement and unwind conditions, reveals the opportunities in this analysis. Dynamics, trade and other, are a recurring theme in this text. While the end result is what makes it to the bank and investor reports, dynamics of how the result is made are critical for identifying problems and opportunities. They are also important to understand from the perspective of explaining monthly return variability to investors when trades extend over calendar month-end boundaries.

Figure 3.9 shows the archetypal trio of trade cumulative return trajectories: (a) gain from trade inception to unwinding; (b) loss from inception to trade cancellation; (c) the J-curve of initial loss followed by recovery and gain. Analysis of collections of trades in each category can reveal possibilities for strategy improvement. Imagine what you would do with the discovery of a distinct characterization of price

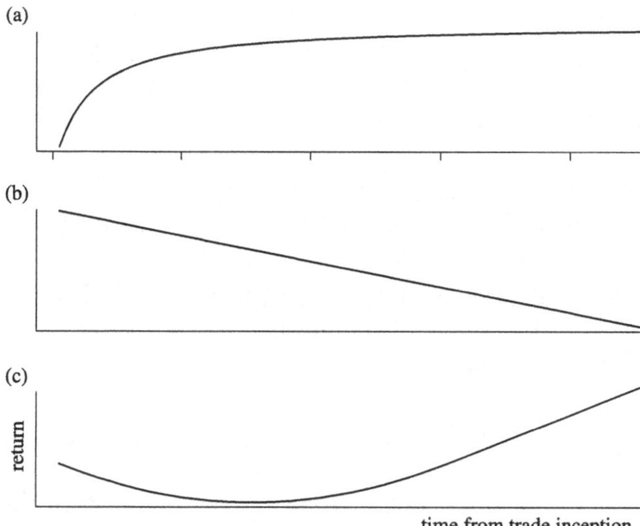

FIGURE 3.9 Archetypal trade cumulative return trajectories

and volume history immediately preceding trade signals that divided prospective trades into the three categories.[3] Imagine.

3.10 FACTOR ANALYSIS PRIMER

The following material is based on the description of factor analysis in *The Advanced Theory of Statistics*, Volume 3, Chapter 43, by Sir Maurice Kendall, Alan Stuart, and Keith Ord (now called *Kendall's Advanced Theory of Statistics [KS]*). The notation is modified from KS so that matrices are represented by capital letters. Thus, γ in KS is Γ here. This makes usage consistent throughout.

Suppose there are p stocks, returns for which are determined linearly from values of $m < p$ unobservable factors:

$$r_j = \sum_{k=1}^{m} l_{jk} f_k + \mu_j + \epsilon_j, \qquad j = 1, \ldots, p$$

[3]This type of research has received considerable attention in seismology where predicting earthquakes remains a research priority for several countries, recently highlighted by the tsunami death toll of over 200,000 from the December 2004 event in the Indian Ocean.

where the ϵs are error terms (observation error, model residual structure). The coefficients l are called *factor loadings*. The variable means μ_j are usually subtracted before analysis. In our case, we assume that returns have mean zero so that $\mu_j = 0$. In matrix form:

$$\underset{(p \times 1)}{r} = \underset{(p \times m)(m \times 1)}{L \quad f} + \underset{(p \times 1)}{\mu} + \underset{(p \times 1)}{\epsilon}$$

where L is the $p \times m$ matrix of coefficients $\{l_{ij}\}$. (Note that this expression is for one set of observations; that is, the set of returns on p stocks for a single day.) Now assume:

1. That the f's are independent normal variables with zero mean and unit variance
2. That each ϵ_j is independent of all other ϵs and of all the fs and has variance (or *specificity*) σ_j^2

It follows that:

$$\text{cov}(r_j, r_k) = \sum_{t=1}^{m} l_{jt} l_{kt}, \qquad j \neq k,$$

$$\text{var}(r_j) = \sum_{t=1}^{m} l_{jt}^2 + \sigma_j^2$$

These relationships may be expressed succinctly in vector/matrix form as:

$$\Gamma = LL' + \Sigma$$

where Σ is the $p \times p$ matrix $\text{diag}(\sigma_1^2, \ldots, \sigma_p^2)$.

From the data, we observe empirical values of Γ. The objectives are to determine the number of factors, m, and to estimate the constants in L and Σ. Determination of m is highly subjective; it is like choosing the number of components in principal component analysis. Indeed, PCA is often used to get an initial estimate of m, which may be refined by likelihood ratio testing and residual analysis of the m-factor model. In what follows, assume that m is fixed.

In some cases, interest is on the implied *factor scores* for particular days in a sample. That is, given returns $r_{,t} = (r_{1,t}, \ldots, r_{p,t})'$ on day t, what are the implied values $f_{,t} = (f_{1,t}, \ldots, f_{m,t})'$ for the m factors? If L and Σ are known, generalized least squares estimation of $f_{,t}$ is obtained by minimizing:

$$(r_{,t} - \mu - Lf_{,t})' \Sigma^{-1} (r_{,t} - \mu - Lf_{,t})$$

Note that the mean stock return vector, μ, is assumed to be zero. (Recall that μ_j is the mean return of stock j; μ is *not* the mean stock return on day t.) The solution of the minimization is:

$$\hat{f}_{,t} = J^{-1} L' \Sigma^{-1} r_{,t}$$

where $J = L'\Sigma^{-1}L$. In practice, L and Σ are unknown; the MLEs are substituted.

An alternative estimator for the factor scores is given in S.J. Press, *Applied Multivariate Analysis*, Section 10.3. (Our notation is used for consistency herein.) Essentially, he assumes away the error covariances when the model is restated as:

$$f_{,t} = Ar_{,t} + u_{,t}, \qquad t = 1, \ldots, n$$

where the factor scores at time t are linear combinations of the stock returns at that time. A subsequent appeal to a large sample approximation results in the estimator:

$$\hat{f}_{,t} = \hat{L}'(nRR')^{-1} r_{,t}$$

3.10.1 Prediction Model for Defactored Returns

In the model described in Section 3.6, interest is in the defactored returns. For day t, the set of defactored stock returns is defined as the difference between the observed set of returns and the weighted factor scores (where the weights are, of course, the factor loadings):

$$dfr_{,t} = r_{,t} - \hat{L}'\hat{f}_{,t}$$

This vector of defactored returns, computed for each day in the sample, provides the raw time series from which the prediction model is constructed. In an autoregressive model, for example, the entry in the regression for day t for stock j is:

$$\sum_{a=1}^{k} dfr_{j,t-a} = \beta_1 dfr_{j,t-k} + \cdots + \beta_q dfr_{j,t-k-q+1} + \epsilon_{j,t}$$

This equation states that the k-day cumulative defactored return to day t is regressed on the q daily defactored returns immediately preceding the cumulation period. Notice that the regression coefficients are common across stocks.

The forecast of the k-day ahead cumulative defactored return at the end of day t is constructed as:

$$\sum_{a=1}^{k} dfr_{j,t+a} = \hat{\beta}_1 dfr_{j,t} + \cdots + \hat{\beta}_q dfr_{j,t-q+1}$$

Other forecast models may be employed: "You pay your money and take your chances."

CHAPTER 4

Law of Reversion

Now here, you see it takes all the running you can do, to keep in the same place.
—*Through the Looking Glass*, Lewis Carroll

4.1 INTRODUCTION

In this chapter, we begin a series of four excursions into the theoretical underpinnings of price movements exploited in statistical arbitrage. The first result, presented in this chapter, is a simple probability theorem that evinces a basic law guaranteeing the presence of reversion in prices in an efficient market. In Chapter 5 a common confusion is cleared up regarding the potential for reversion where price distributions are heavy tailed. In summary, reversion is possible with any source distribution. Following that clarification, we discuss in Chapter 6 definition and measurement of interstock volatility, the variation which is the main course of reversion plays. Finally in this theoretical series, we present in Chapter 7 a theoretical derivation of how much reversion can be expected from trading a pair.

Together these four chapters demonstrate and quantify the opportunity for statistical arbitrage in ideal (not *idealized*) market conditions. The material is not necessary for understanding the remainder of the book, but knowledge of it will amplify appreciation of the impact of market developments that have led to the practical elimination of the discipline of statistical arbitrage in the public domain.

4.2 MODEL AND RESULT

We present a model for forecasting prices of financial instruments that guarantees 75 percent forecasting accuracy. The chosen setting is prediction about the daily spread range of a pair but a little reflection will reveal a much wider applicability. Specifically, we focus on predicting whether the spread tomorrow will be greater or smaller than the spread today.

The model is quite simple. If the spread today is greater than the expected average spread, then predict that the spread tomorrow will be smaller than the spread today. On the other hand, if the spread today was less than the expected average spread, then predict that the spread tomorrow will be greater than the spread today.

4.2.1 The 75 Percent Rule

The model just described is formalized as a probability model as follows. Define a sequence of identically distributed, independent *continuous* random variables $\{P_t, t = 1, 2, \ldots\}$ with support on the nonnegative real line and median m. Then:

$$\Pr[(P_t > P_{t-1} \cap P_{t-1} < m) \cup (P_t < P_{t-1} \cap P_{t-1} > m)] = 0.75$$

In the language of the motivating spread problem, the random quantity P_t is the spread on day t (a nonnegative value), and days are considered to be independent. The two compound events comprising the probability statement are straightforwardly identified with the actions specified in the informal prediction model above. But a word is in order regarding the details of each event. It is crucial to note that each event is a conjunction, *and*, and not a conditional, *given that*, as might initially be considered appropriate to represent the *if–then* nature of the informal model. The informal model is a prescription of the action that will be taken; the probability in which we are interested is the probability of how often those actions (predictions) will be correct. Thus, looking to expected performance, we want to know how often the spread on a given day will exceed the spread on the previous day when *at the same time* the spread on that previous day does not exceed the median value. Similarly, we want to know how often the spread on a given day will not exceed the spread on the previous day when *at the same time* the spread on that previous day exceeds the median.

Those distinctions may seem arcane but proper understanding is critical to the correct evaluation of expected result of a strategy. Suppose that on eight days out of ten the spread is precisely equal to the median. Then the scheme makes a prediction only for 20 percent of the time. That understanding flows directly from the conjunction/disjunction distinction. With the wrong understanding a five-to-one ratio of expected return to actual return of a scheme would ensue.

Operationally, one may bet on the outcome of the spread tomorrow once today's spread is confirmed (close of trading). On those days for which the spread is observed to be greater than the median spread, the bet for tomorrow is that the exhibited spread tomorrow will be less than the spread seen today. The proportion of winning bets in such a scheme is the conditional *given that* probability:

$$\Pr[P_{t+1} < P_t | P_t > m] = \frac{3}{4}$$

Similarly, bets in the other direction will be winners three quarters of the time. Does this mean that we win "1.5 of the time?" Now that really would be a statistical arbitrage! The missing consideration is the relative frequency with which the conditioning event occurs. Now, $P_t < m$ occurs half of the time by definition of the median. Therefore, half of the time we will bet on the spread decreasing relative to today and of those bets, three quarters will be winners. The other half of the time we will bet on the spread increasing relative to today and of those bets, three quarters will also be winners. Thus, over all bets, three quarters will be winners. (In the previous illustration, the conditioning events occur only 20 percent of the time and the result would be $\frac{3}{4} \times \frac{1}{5}$ or just $\frac{3}{20}$.)

Before proceeding to the proof of the result, note that the assumption of continuity is crucial (hence the emphasis in the model statement). It is trivial to show that the result is not true for discrete variables (see the final part of this section).

4.2.2 Proof of the 75 Percent Rule

The proof of the result uses a geometric argument to promote visualization of the problem structure. An added bonus is that one

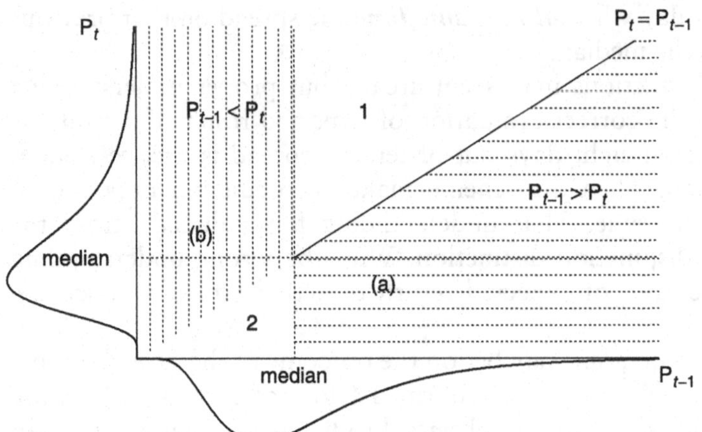

FIGURE 4.1 Domain of joint distribution of P_t and P_{t-1}

can see that certain structural assumptions made in the theorem may be relaxed. These relaxations are discussed following the proof of the basic result.

Consider the joint distribution of two consecutive terms of the sequence, P_{t-1} and P_t. Assuming independence, the contours of this joint distribution are symmetric (about the line $P_t = P_{t-1}$) *regardless of the precise form of the underlying distribution*. In particular, it is *not* necessary to assume that the distribution has a symmetric density function.

Consider Figure 4.1. The domain of the joint distribution (the positive quadrant of $\Re^2$ including the zero boundaries) is partitioned in both dimensions at the median point. By the definition of the median, the four quadrants so constructed each represent 25 percent of the joint distribution.

The lower left and upper right quadrants are bisected radially from the joint median by the axis of symmetry. Now, the symmetry of the density contours—resulting from independent, identical marginal distributions—means that both halves of each quadrant cover the same total probability. Therefore, each half-quadrant accounts for 12.5 percent of the total joint probability.

The remainder of the proof consists of identifying on the figure those regions corresponding to the union in the probability statement made earlier. This is clearly and precisely the union of shaded regions

(a) and (b), which is the domain of the joint distribution excepting unshaded regions (1) and (2). The latter regions each account for 12.5 percent of the total joint probability as shown in the previous paragraph. Therefore, the union of regions (a) and (b) represents exactly three quarters of the joint probability.

It is worth noting at this point that we did not decompose the upper left or lower right quadrants. Fortunately, it is not necessary to do so since there is no specific result for the general case.

4.2.3 Analytic Proof of the 75 Percent Rule

The purpose of giving a geometric argument is to facilitate understanding of generalizations of the result that will be presented in the next section. Before proceeding thereto, we establish the result analytically. Write $X = P_t$ and $Y = P_{t-1}$ to simplify notation. The two events:

$$\{X < Y \cap Y > m\} \quad \text{and} \quad \{X > Y \cap Y < m\}$$

are disjoint (easily seen from the fact that $Y > m$ and $Y < m$ cannot occur simultaneously: On the graph, regions (a) and (b) do not overlap), so the probability of the disjunction is simply the sum of the individual probabilities. Consider the first part of the disjunction:

$$\Pr[X < Y \cap Y > m] = \int_m^\infty \int_{-\infty}^y f_{XY}(x, y)\,dx\,dy$$

where $f_{XY}(x, y)$ denotes the joint density function of X and Y. By the assumption of independence, the joint density is just the product of the individual marginal densities, which in this case are identical (also by assumption). Denoting the marginal density generically by $f(.)$ and its corresponding distribution by $F(.)$, proceed as follows:

$$\begin{aligned}\int_m^\infty \int_{-\infty}^y f_{XY}(x, y)\,dx\,dy &= \int_m^\infty \int_{-\infty}^y f(x)f(y)\,dx\,dy \\ &= \int_m^\infty F(y)f(y)\,dy \\ &= \int_m^\infty F(y)\,dF(y)\end{aligned}$$

The last step is simply recognition that the density function of a random quantity is the analytic derivative of the corresponding distribution function. The remaining steps are trivial:

$$\begin{aligned}\int_m^\infty F(y)dF(y) &= \frac{1}{2}F(y)^2\Big|_m^\infty \\ &= \frac{1}{2}\left[\left(\lim_{t\to\infty} F(t)\right)^2 - F(m)^2\right] \\ &= \frac{1}{2}\left[1 - \left(\frac{1}{2}\right)^2\right] \\ &= \frac{3}{8}\end{aligned}$$

For the second part of the disjunction, the result follows from a similar argument after an initial algebraic simplification. First, note that the event $Y < m$ may be expressed as the union of two disjoint events:

$$Y < m \equiv \{(X > Y) \cap (Y < m)\} \cup \{(X < Y) \cap (Y < m)\}$$

By definition (recall that m is the median of the distribution), the probability of the event $Y < m$ is one half. Therefore, using the fact that probabilities for disjoint events are additive, we may write:

$$\Pr[X > Y \cap Y < m] = \frac{1}{2} - \Pr[X < Y \cap Y < m]$$

Now, proceeding much as for the first part:

$$\begin{aligned}\Pr[X > Y \cap Y < m] &= \frac{1}{2} - \int_{-\infty}^{m}\int_{-\infty}^{y} f_{XY}(x,y)dxdy \\ &= \frac{1}{2} - \int_{-\infty}^{m} F(y)dF(y) \\ &= \frac{1}{2} - \frac{1}{2}\left[F(m)^2 - \left(\lim_{t\to-\infty} F(t)\right)^2\right]\end{aligned}$$

$$= \frac{1}{2} - \frac{1}{2}\left[\left(\frac{1}{2}\right)^2\right]$$

$$= \frac{3}{8}$$

Adding the probabilities of the two parts yields the result.

4.2.4 Discrete Counter

Consider a random variable that takes on just two distinct values, a and b, with probabilities p and $1 - p$, respectively. Whenever $p \neq \frac{1}{2}$, the probability of the random variable exceeding the median is also not equal to one half! Notwithstanding that minor oddity, examine the probabilities of the two events comprising the theorem statement:

$$\{X < Y \cap Y > m\} \quad \text{and} \quad \{X > Y \cap Y < m\}$$

In this discrete example these events are specifically:

$$\{X = a \cap Y = b\} \quad \text{and} \quad \{X = b \cap Y = a\}$$

which each have probability $p(1 - p)$, hence, the probability in the theorem is $2p(1 - p)$. The maximum value this probability can assume is $\frac{1}{2}$ when $p = \frac{1}{2}$ (differentiate, equate to zero, solve).

4.2.5 Generalizations

Financial time series are notorious for the tenacity with which they refuse to reveal underlying mathematical structure (though Mandelbrot, 2004, may demur from that statement). Features of such data, which often show up in statistical modeling, include: nonnormal distributions (returns are frequently characterized by leptokurtosis); nonconstant variance (market dynamics often produce bursts of high and low volatility, and modelers have tried many approaches from GARCH and its variants to Mandelbrot's fractals, see Chapter 3); and serial dependence. The conditions of the theorem can be relaxed to accommodate all of these behaviors.

The result extends to arbitrary continuous random variables directly: The constraint of support on the nonnegative real line is not required. In the geometric argument, no explicit scaling is required for the density axes (the zero origin is convenient for explication). In the analytic argument, recall that we did not restrict the region of support of the densities.

Note that if the underlying distribution has a symmetric density function (implying either that the support is the whole real line or a finite interval), then the pivotal point is the expected value (mean) of the density *if it exists*. The Cauchy distribution, sometimes appropriate for modeling daily price moves, does not have a defined expected value, but it does have a median and the stated result holds.

The 75 percent rule is extended for nonconstant variances in Section 4.3.

The independence assumption is straightforwardly relaxed: From the geometric argument, it is only necessary that the contours of the joint distribution be symmetric. Therefore, the independence condition in the theorem can be replaced by zero correlation. An analytical treatment, with examples, is presented in Section 4.4.

Finally, generalizing the argument for the nonconstant variance case extends the result so that the spread distribution may be different every day, providing that certain frequency conditions are satisfied. Details are given in Section 4.5.

4.3 INHOMOGENEOUS VARIANCES

Spreads are supposed to be generated independently each day from a distribution in a given family of distributions. The family of distributions is fixed but the particular member from which price is generated on a given day is uncertain. Members of the family are distinguished only by the variance. Properties of the realized variance sequence now determine what can be said about the price series.

What can be said if the variances, day to day, exhibit independent "random" values? Then spreads will look as if drawn from, not a member of the original family, but from an average of all the members of the family where the averaging is over the relative frequencies of the possible variances. In other words, the spread on day t is no longer generated from F for a given σ but from the integrated distribution:

$$F_P(p) = \int F_\sigma(p)\mathrm{d}\sigma$$

For example, suppose that the family of variance conditional distributions is the normal family (with constant mean in this context) and that the variances occur randomly as if generated from an inverse gamma distribution; then spreads will look as if they were generated by the Student t distribution. The key to the result is the random element; it guarantees (probabilistically) that the daily transitions look *as if* the underlying spread model is Student t. (This point is expanded upon in Section 4.5 where a similar argument proves the result for arbitrarily different generating distributions day-to-day. An extended discussion of the relationship of marginal distribution to a time series of values is given in Chapter 5.)

We can therefore state that the 75 percent rule is true in the case of inhomogeneous variance sequences.

Note that the distributions for spread (conditional on variance) and for (unconditional) variance need not be of mathematically convenient forms as used in the previous example. Any regular ("well behaved," in terms of continuity) distributions will yield the 75 percent result. There is no requirement for the density or distribution function to be expressed in a closed form mathematical expression.

4.3.1 Volatility Bursts

Autoregressive conditional heteroscedastic (ARCH) models (Engle, 1982) were introduced to capture the observed clustering of variances in macro economic data. In the past few years ARCH and GARCH models have been heavily used in the econometric and finance literature, the latter because of the oft remarked phenomenon of volatility bursts. Most such bursts are of increased volatility from a regular level, typically associated with bad news about a company. Historically, bursts of low volatility are less frequently experienced. Since early 2003, however, volatility of stocks on the U.S. exchanges has been declining. Spread volatility reached unprecedented lows in 2003 and 2004; implications of that development for statistical arbitrage are examined in Chapter 9.

When volatility exhibits bursts, variances are not generated independently each day but exhibit serial correlation. The 75 percent rule still holds by this argument: Within a burst, the theorem applies as if the situation were (approximately) constant variance. Therefore, only the transition days could alter the result. There are comparatively few such days, so the result will stand to a very close approximation. In fact, the result can be shown to hold exactly: The transitions are irrelevant—see the argument in Section 4.5 for the general nonconstant variance case. Chapter 5 presents analysis of related patterns.

4.3.2 Numerical Illustration

Figure 4.2(a) shows the histogram of a sample of 1,000 values generated from the normal–inverse Gamma scheme:

$$\sigma_t^2 \sim \text{IG}[a, b],$$
$$P_t \sim \text{N}[0, \sigma_t^2]$$

First, generate an independent drawing of σ_t^2 from the inverse Gamma distribution (with parameters a and b—the actual specification of a

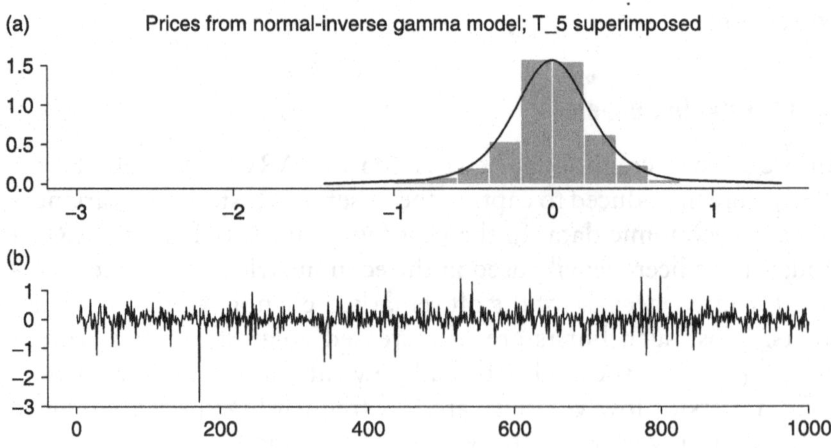

FIGURE 4.2 Random sample from normal–inverse Gamma model

and b does not matter: Any nonnegative values will do). Then, using this value of σ_t^2, generate a value for P_t from the normal distribution with mean 0 and variance σ_t^2. Superimposed on the histogram is the density function of the Student t distribution, which is the theoretical marginal distribution of spreads here. Figure 4.2(b) shows the sample as a time series.

The proportion of one-day moves in the direction of the median is 75.37 percent, satisfyingly in accord with the rule.

4.4 FIRST-ORDER SERIAL CORRELATION

The result can be extended to the case of correlated variables. The simplest case to consider is that of distributions with symmetric density functions, since then the contours are circles (uncorrelated) or ellipses (correlated). In the latter case, one can see that by dividing up $\Re^2$ into quadrants along the major and minor axes of the contours, then bisecting those quadrants radially from the joint median point as previously, one is left with equiprobable regions once again. (Recall that, with symmetric densities, all quadrants are probablistically bisected this way, not just those corresponding to the lower left and upper right in the rotated coordinates.) The remaining task is to identify the half quadrants with a correct statement (like the one with which the original result was stated) in terms of the random quantities. The result is easily seen by example. Suppose that P_t and P_{t-1} have covariance c. Define a new variable as a linear combination of the correlated variables, $Z_t = a(P_t - rP_{t-1})$. The coefficient r is set to:

$$r = \frac{cov[P_t, P_{t-1}]}{var[P_{t-1}]}$$

(which is just the correlation between P_t and P_{t-1}) in order to make P_{t-1} and Z_t uncorrelated; the scale factor a is chosen to make the variance of Z_t equal to the variance of P_t:

$$a = (1 - r^2)^{-\frac{1}{2}}$$

Now the theorem applies to P_{t-1} and Z_t *providing that Z_t has the same distribution as P_{t-1}*, so that we have:

$$\Pr[(Z_t < P_{t-1} \cap P_{t-1} > m) \cup (Z_t > P_{t-1} \cap P_{t-1} < m)] = 0.75$$

Substituting for Z_t converts the expression into a form involving the original variables:

$$\Pr[(aP_t - arP_{t-1} > P_{t-1} \cap P_{t-1} > m) \\ \cup (aP_t - arP_{t-1} > P_{t-1} \cap P_{t-1} < m)] = 0.75$$

Rearrangement of terms gives the required expression:

$$\Pr[(P_t < (a^{-1} + r)P_{t-1} \cap P_{t-1} > m) \\ \cup (P_t > (a^{-1} + r)P_{t-1} \cap P_{t-1} < m)] = 0.75$$

Clearly, the case of zero correlation, equivalently $r = 0$ and $a = 1$, with which we began is a special case of this more general result.

The boundary, $P_t = (a^{-1} + r)P_{t-1}$, partitions the quadrants of $\Re^2$ in proportions determined by the size of the correlation. In the uncorrelated case, the quadrants are bisected as we saw earlier. Figure 4.3 shows the relationship of $a^{-1} + r = \sqrt{1 - r^2} + r$ with r. The maximum, $\sqrt{2}$, occurs at $r = \sqrt{\frac{1}{2}}$ (easily shown analytically by the usual procedure of differentiating, equating to zero, and solving).

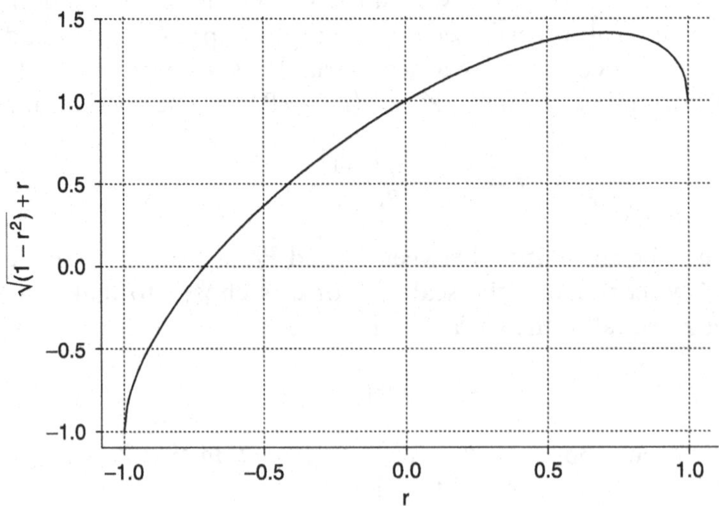

FIGURE 4.3 r versus $\sqrt{(1 - r^2)} + r$

It is important not to lose sight of the additional constraint introduced in the argument here. The theorem applies to correlated variables providing that a linear combination of variables with the stated marginal distribution retains that distributional form. This is true of normal variables, bivariate Student t variables, and lots of others. But it is not true in general.

At the limit when P_t and P_{t-1} are perfectly correlated ($r \to 1, a \to \infty$) the result breaks down. Failure is caused by the singularity as the original two degrees of freedom (two distinct days or observations) collapse into a single degree of freedom (two days constrained to have the same price so the reversion statement of the theorem is impossible).

4.4.1 Analytic Proof

The frequency of one-day moves in the direction of the median is given by the probability:

$$\Pr[(P_t < P_{t-1} \cap P_{t-1} > m) \cup (P_t > P_{t-1} \cap P_{t-1} < m)]$$

Consider the first part of the disjunction:

$$\begin{aligned}
\Pr[P_t < P_{t-1} \cap P_{t-1} > m] &= \int_m^\infty \Pr[P_t < P_{t-1} \cap P_{t-1} = p]\mathrm{d}p \\
&= \int_m^\infty \Pr[P_t < P_{t-1} | P_{t-1} = p]\Pr[P_{t-1} = p]\mathrm{d}p \\
&= \int_m^\infty \Pr[P_t < p | P_{t-1} = p]\Pr[P_{t-1} = p]\mathrm{d}p
\end{aligned}$$

Notation is abused here to emphasize the logic. For continuous quantities, it is not correct to write $\Pr[P_{t-1} = p]$ since the probability of the quantity taking on any specific value is zero. The correct expression is the density function evaluated at p:

$$\Pr[P_t < P_{t-1} \cap P_{t-1} > m] = \int_m^\infty \Pr[P_t < p | P_{t-1} = p] f(p)\mathrm{d}p$$

Note that the conditional cumulative probability (first term) reduces to the unconditional value $\Pr[P_t < p]$ when P_t and P_{t-1} are independent, the case considered in Section 4.2.

In order to simplify notation in the remaining derivation of the result, let X denote P_t and Y denote P_{t-1}. Then the probability of interest is:

$$\Pr[X < Y \cap Y > m] = \int_m^{\infty} \Pr[X < p | Y = p] f(p) dp$$

Expand the conditional cumulative probability $Pr[X < p|Y = p]$ into the integral of the conditional density to obtain $f_{X|Y}(p) = Pr[X = x|Y = p]$:

$$\begin{aligned}
\Pr[X < Y \cap Y > m] &= \int_m^{\infty} \int_{-\infty}^{p} f_{X|Y}(x) dx f(p) dp \\
&= \int_m^{\infty} \int_{-\infty}^{p} f_{X|Y}(x) f(p) dx dp \\
&= \int_m^{\infty} \int_{-\infty}^{p} f_{XY}(x, p) dx dp
\end{aligned}$$

where $f_{XY}(\ldots)$ denotes the joint density function of X and Y.

Now, using:

$$\int_m^{\infty} \int_{-\infty}^{\infty} f_{XY}(x,p) dx dp = \frac{1}{2}$$

(since the inner integral reduces to the marginal density of X and, by definition of the median, the outer integral is then precisely one half), and noting that:

$$\int_m^{\infty} \int_{-\infty}^{\infty} f_{XY}(x, p) dx dp = \int_m^{\infty} \int_{-\infty}^{p} f_{XY}(x, p) dx dp + \int_m^{\infty} \int_p^{\infty} f_{XY}(x, p) dx dp$$

it follows immediately that:

$$\int_m^{\infty} \int_{-\infty}^{p} f_{XY}(x, p) dx dp = \frac{1}{2} - \int_m^{\infty} \int_p^{\infty} f_{XY}(x, p) dx dp$$

This may seem like an irrelevant diversion but, in fact, it takes the proof to within two steps of completion. At this point, we invoke the symmetry of the joint density (which follows from the assumption of identical marginal distributions). Formally, an expression of symmetry is:

$$\int_m^\infty \int_m^p f_{XY}(x,p)\mathrm{d}x\mathrm{d}p = \int_m^\infty \int_m^x f_{XY}(x,p)\mathrm{d}p\mathrm{d}x$$

Now, reversing the order of integration (be careful to watch the integral limits) yields the algebraic equivalence:

$$\int_m^\infty \int_m^x f_{XY}(x,p)\mathrm{d}p\mathrm{d}x = \int_m^\infty \int_p^\infty f_{XY}(x,p)\mathrm{d}x\mathrm{d}p$$

Therefore:

$$\int_m^\infty \int_m^p f_{XY}(x,p)\mathrm{d}x\mathrm{d}p = \int_m^\infty \int_p^\infty f_{XY}(x,p)\mathrm{d}x\mathrm{d}p$$

Penultimately, note that the sum of the latter two integrals is one quarter (again, by definition of the median):

$$\begin{aligned}\int_m^\infty \int_m^p f_{XY}(x,p)\mathrm{d}x\mathrm{d}p &+ \int_m^\infty \int_p^\infty f_{XY}(x,p)\mathrm{d}x\mathrm{d}p \\ &= \int_m^\infty \int_m^\infty f_{XY}(x,p)\mathrm{d}x\mathrm{d}p = \frac{1}{4}\end{aligned}$$

And so:

$$\begin{aligned}\int_m^\infty \int_{-\infty}^p f_{XY}(x,p)\mathrm{d}x\mathrm{d}p &= \frac{1}{2} - \int_m^\infty \int_p^\infty f_{XY}(x,p)\mathrm{d}x\mathrm{d}p \\ &= \frac{1}{2} - \frac{1}{2}\left(\frac{1}{4}\right) \\ &= \frac{3}{8}\end{aligned}$$

The argument is similar for the second part of the disjunction.

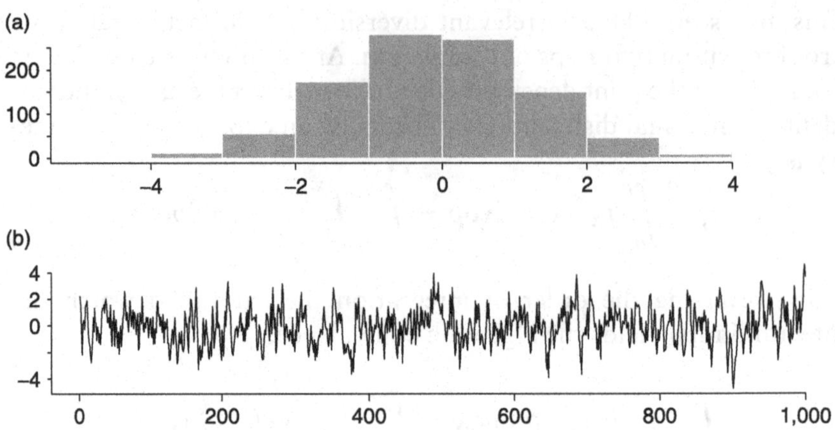

FIGURE 4.4 Random sample from autocorrelated model

4.4.2 Examples

Example 1 One thousand terms were generated from a first-order autoregressive model with serial correlation parameter $r = 0.71$ (see Figure 4.3 and the final remarks in Section 4.4 regarding this choice) and normally distributed random terms. Figure 4.4(b) shows the time plot; Figure 4.4(a) shows the sample marginal distribution.

The proportion of reversionary moves exhibited by the series is 62 percent.

Adding a little more realism, we compute an estimate of the median treating the series as if it were observed day by day. Analysis of the local median adjusted series (using window length of 10) is illustrated in Figure 4.5. A slightly greater proportion of reversionary moves is exhibited by the adjusted series, 65 percent.

4.5 NONCONSTANT DISTRIBUTIONS

Suppose that spreads are generated from a normal distribution for 100 days, followed by a uniform distribution for 50 days. Within each period the basic theorem is applicable; therefore, with one exception in 150 days the 75 percent rule is true. Suppose that on the one hundred fifty-first day, price range is generated from the normal distribution once again. What can we say?

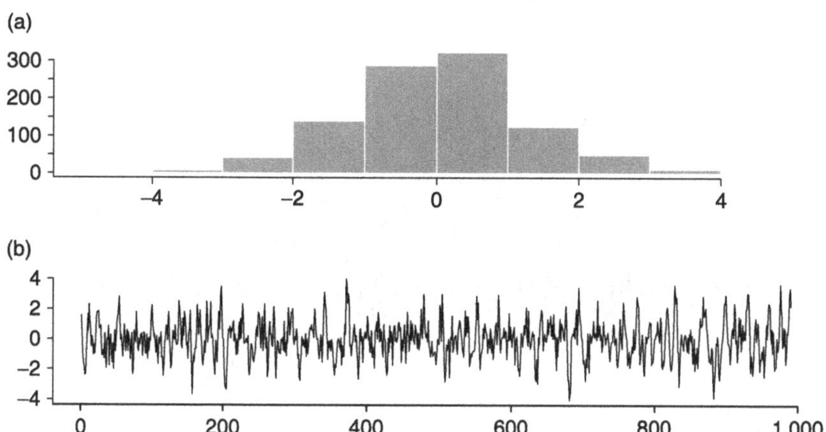

FIGURE 4.5 Random sample from autocorrelated model, locally median adjusted: (a) histogram (b) time series

Unequivocally we can say that the 75 percent rule is true on average throughout the series. The crux of the proof is the two transition days: (1) from normal to uniform, and (2) from uniform to normal. Recall Figure 4.2. Region 1 is probabilistically bounded by $0 < \Pr[(1)] < \frac{1}{4}$ for random quantities from any two continuous, independent distributions (by which it is meant that the probability $\Pr[(1)] = \Pr[(P_t, P_{t-1}) \in (1)]$). This follows from the definition of the median as stated in Section 4.2. Denote this probability by p. Now, the complement of region 1 in the quadrant therefore has probability $\frac{1}{4} - p$ (with the same kind of meaning attached to "probability of a region" as previously). The transitions are the key because only when the distribution changes is the basic result in question. Indeed, it is not hard to show that the result does not hold. For each transition where $p > \frac{1}{8}$ (normal to uniform is a case in point) the theorem result is not 75 percent but $100(1 - 2p)\% < 75\%$. However, for each such transition, the reverse transition exhibits the complementary probability $\frac{1}{4} - p$ for region 1.

Similar analysis applies to region 2. And this is true irrespective of whether the probability of region 2 is the same as the probability of region 1—which it is not if one or the other of the distributions is asymmetric.

Thus, if transitions occur in pairs, the exhibited probability is the average and, hence, the 75 percent result resurfaces. (If both densities

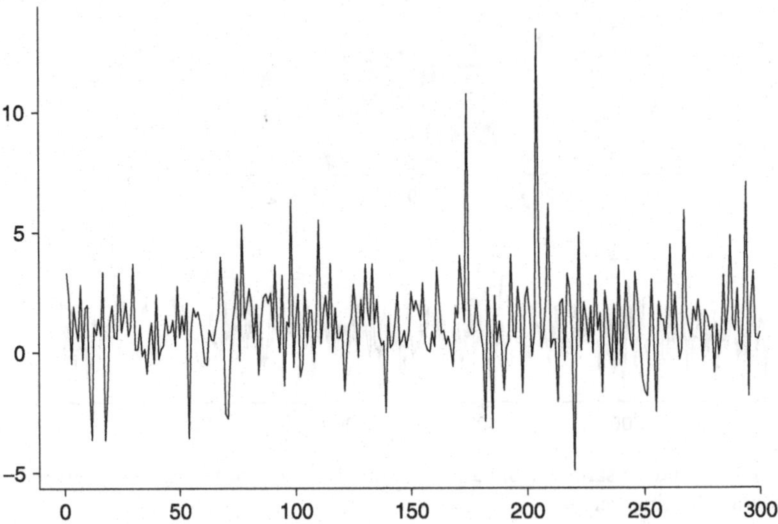

FIGURE 4.6 Random sample from mixed normal, lognormal, and Student t distributions

are symmetric, then the "pairs" condition is sufficient. However, if at least one density is asymmetric, so that $\Pr[(1)] \neq \Pr[(2)]$, then the pairs must occur often enough to statistically guarantee that there are few pairs in region 1–2 compared to 1–1 or 2–2.)

One can push the argument further to prove the case for three alternative distributions, the key caveat being that each of the three distinct pairwise transitions occur equally often in both directions. An appeal to mathematical induction then completes the proof for an arbitrary number of alternative distributions.

4.6 APPLICABILITY OF THE RESULT

After several pages of theoretical development, it is a good idea to pause and ask, "What is the relevance to model-based stock trading?" A major starting assumption is stationarity—a conveniently unaddressed, unmentioned in fact, thus far. We required spreads to be "independent, identically distributed" (later relaxing the conditions to allow for serial correlation); implicit therein in a time series context is stationarity.

Now stock price series are not stationary. What about spreads? They are typically not stationary either. However, dynamically adjusted for local estimates of location and variance, a reasonable approximation to stationarity can be made. There is a link here to the idea of cointegration (see Chapter 3). It may be difficult to uncover structure in individual price series but the difference between series (spreads) more readily yields predictable relationships. Extending the basic notion of cointegration to locally defined series (we might use the nomenclature "locally cointegrated") identifies the link.

It is in this spirit of informed, dynamic approximation that the theoretical result has guiding validity.

4.7 APPLICATION TO U.S. BOND FUTURES

The theorem presented in this chapter, while motivated by the discussion of spreads between stock prices of similar companies (the classic pair of a pairs trade), is applicable with much greater generality. As long as the conditions on the "price" series are reasonably met, the theorem applies to any financial instrument. Of course, the rub is in the meeting of the conditions—many series do not (without more attention to the structural development over time—trend developments for example). Bond prices do show a good fit to the theorem.

U.S. 30-year Treasury bond futures were studied with the simple forecasting model for changes in daily high–low price range. The front future contract, being the most liquid, was examined. (Because of concern about possible distortions arising from contract expirations, the study was repeated with contract rollover at 15 business days prior to expiration of the front future. No distortions were observed in the analysis, hence the results of the vanilla series are reported.) Figure 4.7 shows the sample distribution of the data for 1990–1994 used in the study—a strong skew is evident in the distribution. A time plot of the series is given in Figure 4.8.

In the prediction model, the median value was estimated each day using data for the preceding 20 business days. Operationally, this local median calculation is desirable to minimize the effects of evolutionary change. One of the benefits is a reduction in serial correlation: The raw series (equivalent to using a constant median)

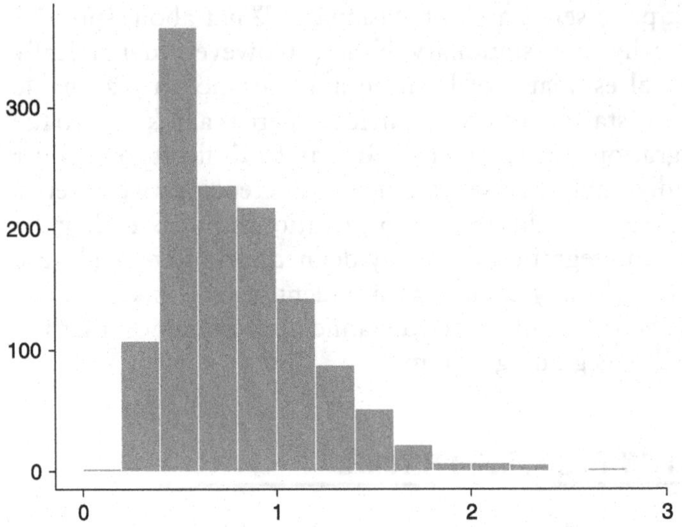

FIGURE 4.7 Marginal distribution of high–low range of front U.S. 30-year bond future

TABLE 4.1 Empirical study of U.S. 30-year bonds

Year	Proportion $P_t > P_{t-1} \mid P_{t-1} < m$	Proportion $P_t < P_{t-1} \mid P_{t-1} > m$	Proportion Overall
1990	70%	75%	73%
1991	77%	72%	74%
1992	78%	76%	77%
1993	78%	76%	77%
1994	78%	76%	77%
All	77%	75%	76%

Note: 250 trading days per year

exhibits autocorrelations in the range [0.15, 0.2] for many lags; the local median adjusted series exhibits no significant autocorrelations.

Results of the forecasting exercise are presented in Table 4.1: They are quite consistent with the theorem.

The result confirmed by bond future prices is economically exploitable. Some sophistication is required in the nature of the implementation of the trading rule, particularly with respect to managing trading costs, but there are many possibilities.

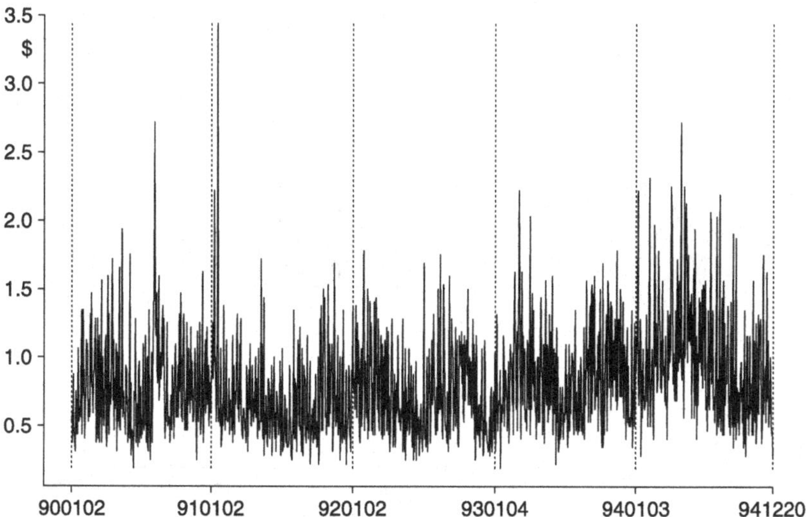

FIGURE 4.8 Daily high–low range of front U.S. 30-year bond future

4.8 SUMMARY

The implication of the theorem statement is a little provocative: The 75 percent forecast accuracy is guaranteed only if the conditions of the theorem are met. In practice, the evidence is that many stock prices and spreads do meet the conditions approximately over short periods. Moreover, the rate of change, when change occurs, is often sufficiently slow that a properly calibrated, dynamic model (local characterization of the mean in the examples examined in this chapter) exhibits reversion results similar to the theoretical prediction.

Empirical evidence for U.S. 30-year Treasury bonds suggests that this market, also, comes close to meeting the conditions. Certainly the accuracy of the empirical model is not demonstrably different from 75 percent for the five years 1990–1994. With a little ingenuity, many situations seemingly violating the conditions of the theorem can be made to approximate them quite closely.

Appendix 4.1: LOOKING SEVERAL DAYS AHEAD

Assuming no persistent directional movement (trending), as we have been doing in the theoretical development and proxying by local

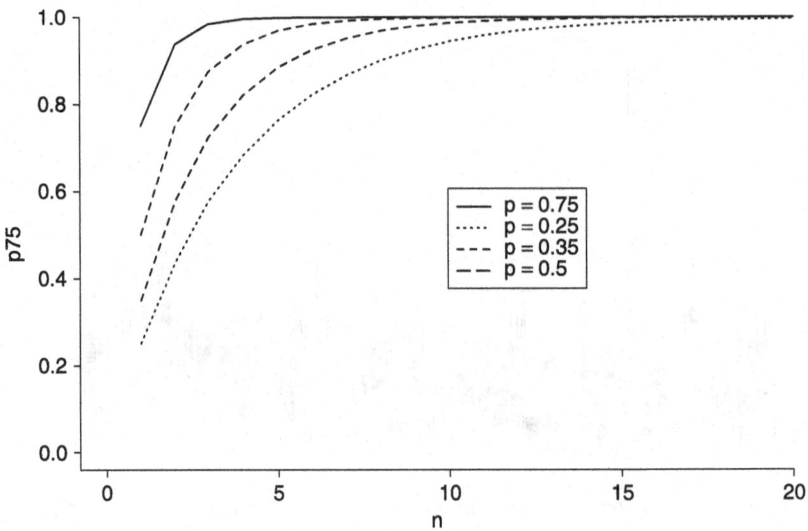

FIGURE 4.9 Probability of at least 1 "winning move" in next n days

location adjustment in the applications, there is obviously greater opportunity for reversion if more than one day ahead may be considered. Of course, it is crucial that *each day* in the k-day period ahead can be looked at individually; if one may only look at k-day movements, then the situation is essentially unchanged from the one-day case. When each day may be considered, there are multiple (independent) chances of a win (a reversion), so the probability of a win increases.

Figure 4.9 shows the probability of a move from the current position in the direction of the median occurring within the next k days for $k = 1, \ldots, 20$. The probability of such a move occurring in one day is 0.75 from the theorem. The other probabilities are calculated from the binomial distribution as described next. The figure also includes graphs assuming a winning probability of 0.25, 0.35, and 0.5: In practice, a lower value than 0.75 will be observed under the best circumstances because of theorem assumption violations.

By assumption, prices are independent each day. The theorem states that the probability of a reversionary move from the current position is 0.75 for *any* day. So, the probability of a reversionary move from the price on day t is 0.75 for day $t+1$, for day $t+2$

regardless of what happens on day $t+1$ (this is the independence assumption), and so on. Thus, the number of days in the next k days that show reversion from today's price is a binomial quantity with parameters k (the number of trials) and 0.75 (the probability of success on a given trial). The probability shown in the graph is then:

$$\begin{aligned}\Pr[1 \text{ or more successes in } k \text{ days}] &= 1 - \Pr[0 \text{ successes in } k \text{ days}]\\ &= 1 - \binom{k}{0} 0.75^0 (1-0.75)^{k-0}\\ &= 1 - 0.25^k\end{aligned}$$

When the independence constraint is relaxed, the binomial result is not valid, deviation from it being a function of the temporal structure of the series. Where that structure is simple trending, the longer ahead one looks the less accurate the binomial probability is. Accuracy may be restored by appropriate attention to the structural form of the prediction function: adding an estimated trend component.

CHAPTER 5

Gauss Is Not the God of Reversion

It is better to be roughly right than precisely wrong.

—J.M. Keynes

5.1 INTRODUCTION

We begin with two quotes:

> *The distribution of nominal interest rates indicates that there is no mean reversion in interest rates and the structure does not resemble a normal distribution.*
>
> *In contrast, real interest rates appear to be normally distributed. The distribution suggests interest rates have the characteristic of mean reversion.*

Both of these quotes, from research publications (several years ago) of a large, successful Wall Street firm, contain fallacious logical implications. There is no explicit definition of what is meant by "mean reversion" and one might, therefore, take the depictions of reversion through marginal distributions of yields as defining it implicitly. But that would mean that the author is not talking about mean reversion in the general sense in which it is widely interpreted.

The first statement is wrong because time series that exhibit marginal distributions of any shape whatsoever can be mean reverting. The 75 Percent Theorem in Chapter 4 proves that unambiguously. Note the caveat: *can be*. It is not sufficient to look at the marginal distribution of a time series in order to be able to correctly make a pronouncement on whether that series exhibits the quality of mean reversion. That is why the second statement is also wrong. It is entirely possible for a time series to be as far from mean reverting as imaginable (well, almost) while at the same time exhibiting a normal marginal distribution, as is demonstrated in the final section of this chapter.

Mean reversion by definition involves temporal dynamics: The sequence in which data, samples from the marginal distribution, occur is the critical factor.

5.2 CAMELS AND DROMEDARIES

The report mentioned at the opening of this chapter shows a histogram of the ten-year bond yields since 1953, presented in a section called, "A Double Humped Distribution." The claim is that a time series exhibiting such an obviously nonnormal marginal distribution cannot be reversionary. Despite the imagery of the title, there are, in fact, several modes in the histogram. However, it is sufficient to demonstrate a reversionary time series that has an extremely pronounced bimodal distribution.

Figure 5.1 shows a combined random sample of 500 values from the normal distribution with mean 4.5 and 500 values from the normal distribution with mean 8.5 (and unit standard deviation for both). These locations are motivated by the major modes of the bond yield distribution; they are not critical to the demonstration.

Figure 5.2 shows one possible time series of these 1,000 points. How much reversion is exhibited, from day 1, by this series? A lot. Consider each segment separately. Take any point in either segment far from the segment average (4.5 in the first segment, 8.5 in the second). How many points must be passed before the time series exhibits a value closer to the segment average? Typically only one or two. The series never departs from the segment mean never to return; on the contrary, the series continually crosses the average—almost a definition of reversion. A veritable excited popcorn process!

That leaves one to consider the implications of the single change point: the move from the first segment to the second in which the segment mean increases. Does a single mean shift in a time series destroy the property of mean reversion? Only if the process has to be reverting to a global mean—that would be an unusually restrictive interpretation of mean reversion; it would also be unhelpful and

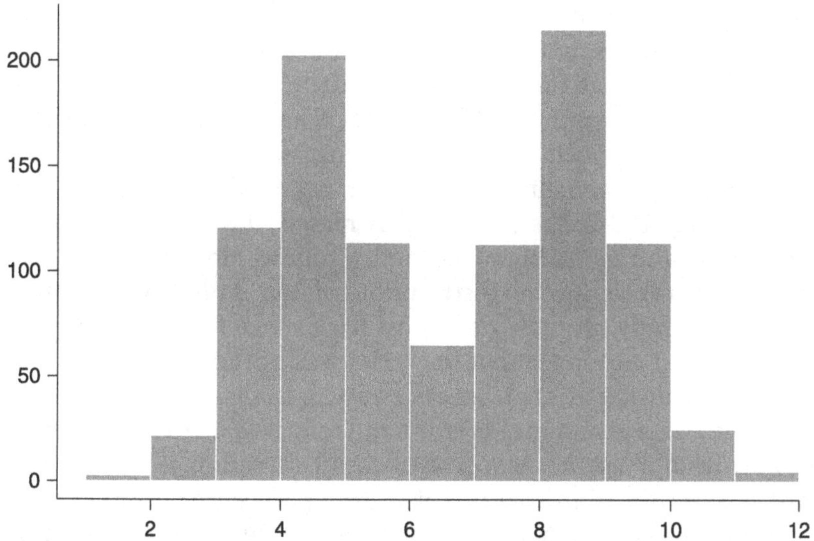

FIGURE 5.1 Marginal distribution: mixture distribution of half N[4.5,1] and half N[8.5,1]

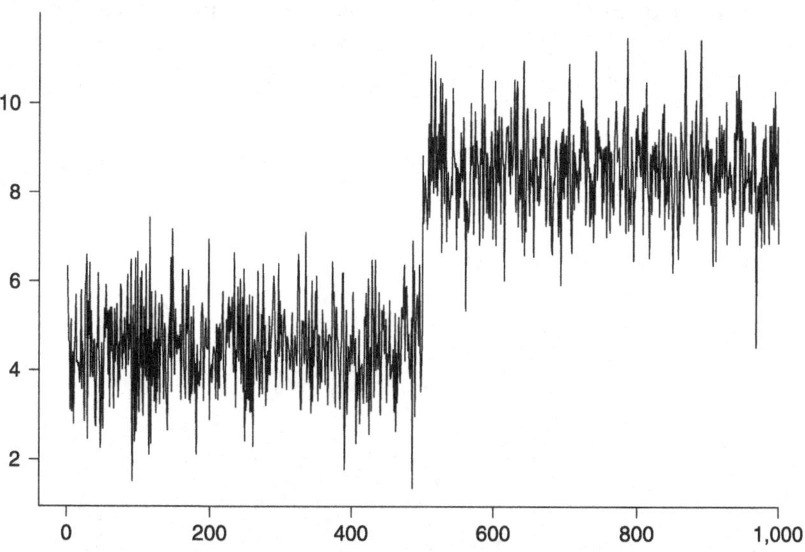

FIGURE 5.2 Time series realization of the sample depicted in Figure 5.1

misleading in a time series context, and misleading, too, in assessing trading opportunities that necessarily occur in a time series, sequential manner. Suppose that the raw data are daily values. Then for two years, the first segment of the combined series, the data would unarguably be described as mean reverting. Shortly after the mean increase the same conclusion would be acknowledged, albeit the mean to which the series reverts is increased. Two more years of splendid reversion to the (new) mean then follow. How could anyone then look at the marginal distribution of the daily data, having traded wonderfully profitably the most basic mean reverting strategy describable, and announce that the series was not mean reverting?

A second distributional concern raised is that the bond data distribution has a pronounced right-hand tail. This feature is actually irrelevant insofar as the property of mean reversion is concerned. Figure 5.3 shows a histogram of 1,200 points: the 1,000 from the normal mixture of the previous example with 200 uniformly randomly distributed over the interval [10,14]. Figure 5.4 shows one possible time series of the 1,200 points: How much reversion is there in this series? A lot, once again. The first two segments are mean reverting as demonstrated in the previous paragraphs. What about the final segment? The points are a random sample from a uniform

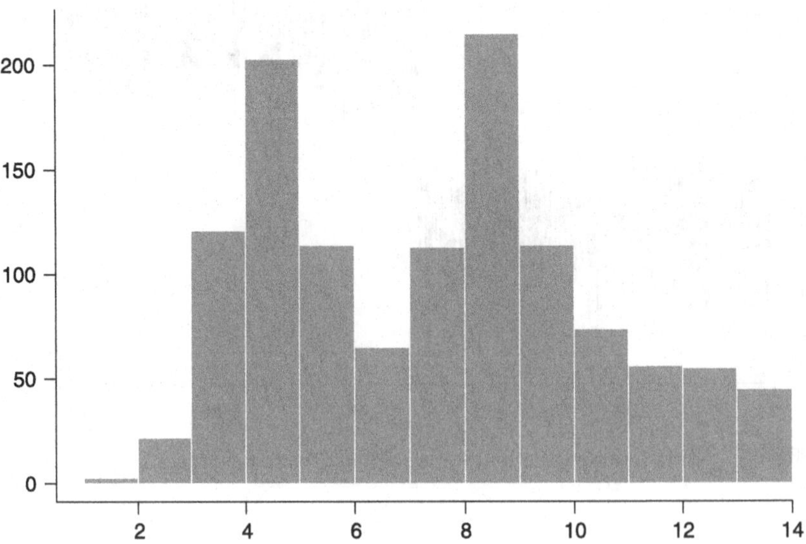

FIGURE 5.3 Marginal distribution with heavy right tail

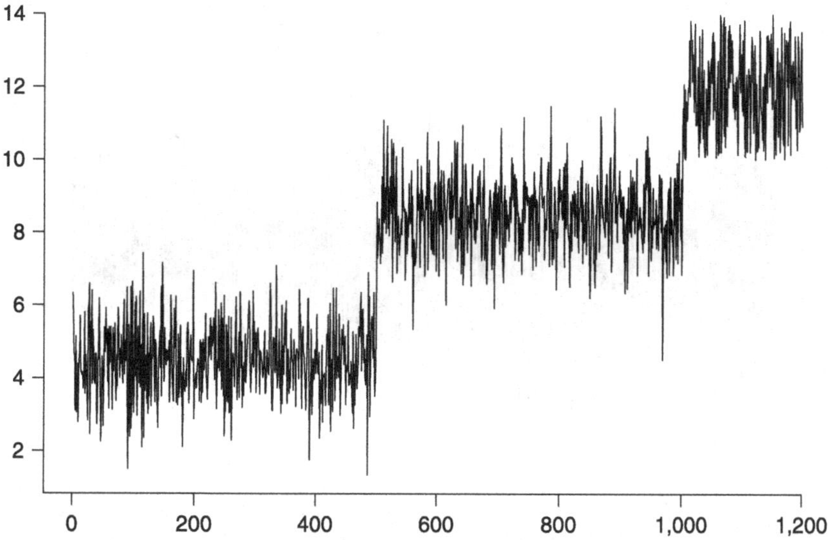

FIGURE 5.4 Time series realization of the sample depicted in Figure 5.3 ordered by underlying sampling distribution

distribution, not a normal distribution, and therefore according to the report, the time series cannot be mean reverting. From the graph, would you agree? If you do, I would like to entertain some gambles with you!

One might reasonably charge that these examples are artificial. Real time series don't exhibit such convenient segmentation. This is undeniable. But it is also not an issue. Figure 5.5 shows another possible realization of the random sample in Figure 5.3. It was obtained by randomly selecting, without replacement, the values in the original sample of 1,200 and plotting in order of selection. How much reversion is exhibited? Lots. Don't you agree just from an eyeball analysis?

Whether the marginal distribution of a time series is a dromedary or a camel really doesn't matter as far as mean reversion is concerned. To repeat: Temporal structure is the critical feature.

5.2.1 Dry River Flow

Camels are distinguished largely by their remarkable adaptation to life in arid regions, the twin key abilities bestowed by evolution being the sponge-like capacity of the body to absorb water and the

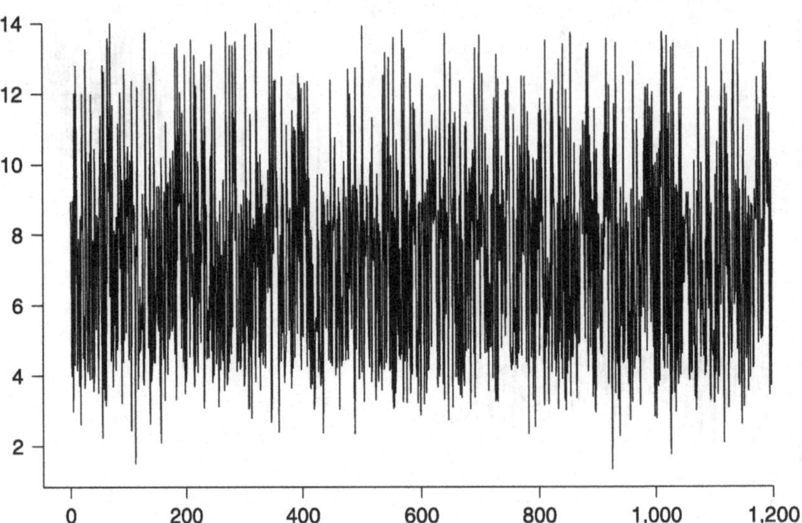

FIGURE 5.5 Time series realization of the sample depicted in Figure 5.3 randomly reordered

drip-feed use thereof. Dry rivers, those with notable periods of no stream flow, are common in arid landscapes (and often the source of replenishment for camels). Looking at a time series of stream flow for a dry river, one would find it difficult to deny the claim that such a series is reversionary. The series always returns to zero no matter how far it departs therefrom in the rainy season.

What is the typical marginal distribution of dry river flow? Obviously it is very asymmetric. So, entirely without recourse to mathematical formalism or rigor, we have a proof that a reversionary time series need not be characterized by a symmetric marginal distribution such as the normal.

The Bell Still Tolls Often a so-called Gamma distribution is a good approximation (allowing for reasonable fudge for the zeroes) to dry river flow. Now, a squared Gaussian variable is distributed as a Chi-squared variable, which is also a Gamma variable. More than curious coincidence?

But Not for Reversion The dry river flow example is just one particular case of self-evident reversion. To repeat: *any* marginal distribution can be exhibited by a reversionary time series.

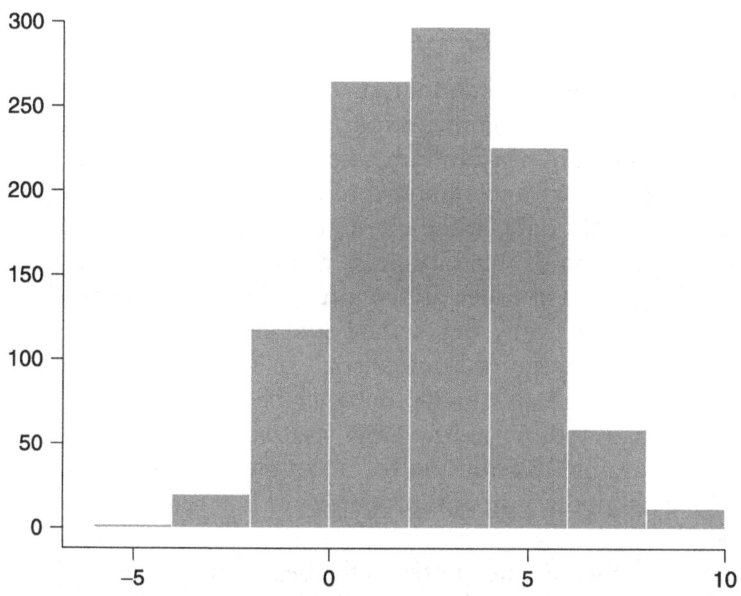

FIGURE 5.6 Random sample of 1,000 points from N[2.65, 2.44^2]

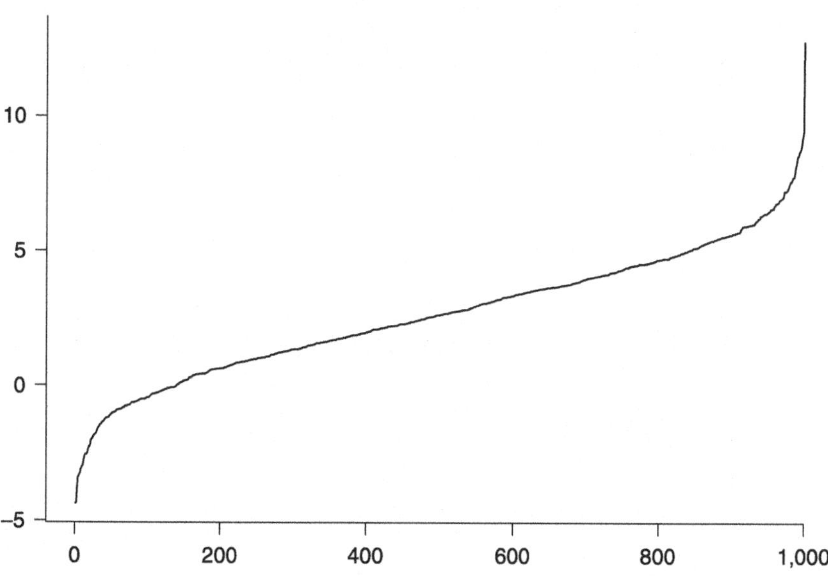

FIGURE 5.7 Time series realization of the sample depicted in Figure 5.6

5.3 SOME BELLS CLANG

A time series exhibiting a normal marginal distribution is not predisposed to exhibit mean reversion. Extraordinary diversion is equally possible. Figure 5.6 shows a random sample from the normal distribution with mean 2.65 and standard deviation 2.44—the sample values of the real yield data in the cited analysis. Figure 5.7 shows one possible realization of this sample as a time series in which the sample points are taken in order of magnitude. How much reversion is exhibited, from day 1, by this series? Not one bit. As with my earlier example, the charge of "unrealistic" or "points don't occur in [size] order" can be readily made. But such charges are irrelevant: The point demonstrated, no matter how stylized the illustration, is that a normal marginal distribution for a sample from a time series reveals nothing about the temporal properties of that series, reversion or any other.

The paper from which the quotes at the beginning of this chapter are taken, and the content of which is critically examined here, is not included in the references list.

CHAPTER 6

Interstock Volatility

. . .the investor does (or should) consider expected return a desirable thing and variance of return an undesirable thing.

—Harry Markowitz, *Journal of Finance*, Vol. 7, No. 1, 1952

6.1 INTRODUCTION

The reversion exploited in pairs trading is the reversion of stock prices to each other following a movement apart, or a movement apart following an unusual narrowing—the popcorn process of Chapter 2. The amount of movement in stock prices is measured by and expressed as stock volatility, which is the standard deviation of price changes (returns). The volatility that is relevant to the spread reversion scheme is the volatility of relative price movements between stocks, hence *interstock* volatility. Figure 6.1(a) shows the daily closing price (adjusted for splits and dividends) of two related stocks, ENL and RUK, for the first six months of 1999. The price series track each other very closely, further demonstrated by the price difference (spread) series shown in Figure 6.1(b). Ignoring the scales, the spread series looks like any stock price series, and it will not be surprising to discover that the volatility of the spread is similar in character to stock price volatility.

Stock price volatility is the standard deviation of returns. But what is the relevant measure of spread volatility? In Chapter 2 we calibrated trade rules by directly computing the standard deviation of the spread itself. Here we are interested in the technical definition

(a) Adjusted close price

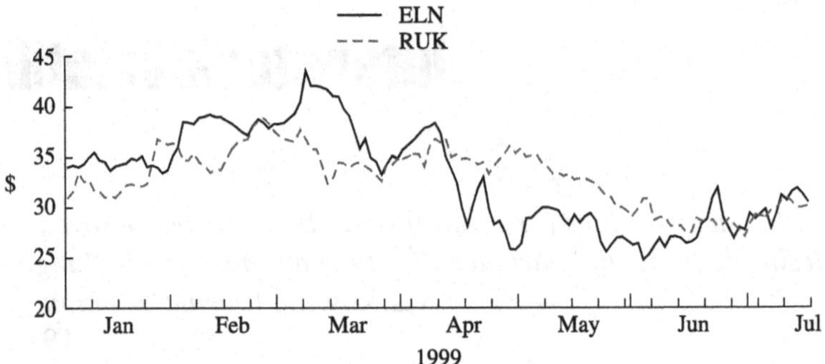

(b) Adjusted close price difference

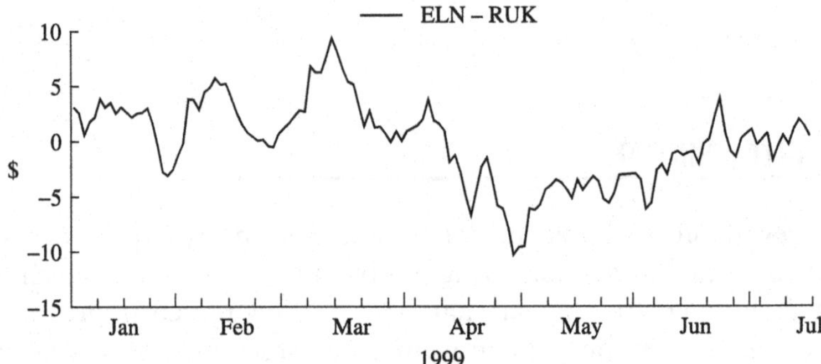

(c) Annualized volatility

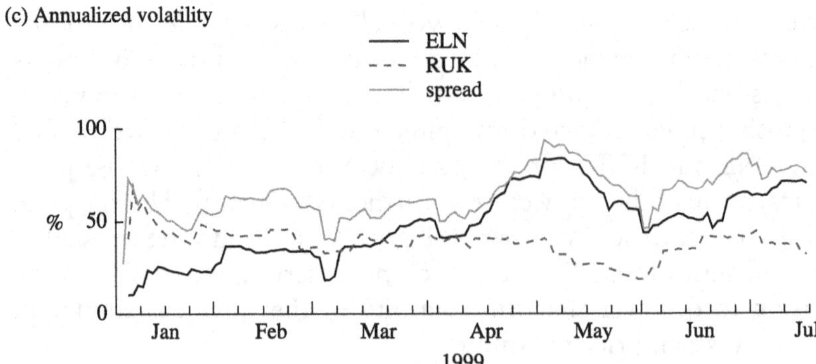

FIGURE 6.1 (a) Daily adjusted close prices of ENL and RUK; (b) spread; (c) volatilities

of volatility, not simply a scale factor, and this requires focusing on an appropriate *return* series. Considering the price difference in Figure 6.1(b), with the trace variously above and below zero, it is obvious that one should not treat the spread as a price—infinite "returns" are readily exhibited by such a series! The relevant measure is apparent from consideration of how a basic reversion strategy exploits spreads: When the spread widens or narrows beyond some threshold whereupon reversion is subsequently expected, make two bets, one on each stock, one a buy and the other a sell. Thus, the spread bet return is the sum of the return on the stock bought and the negative return on the stock sold:

$$\textit{spread return} = \textit{return on buy} - \textit{return on sell}$$

(assuming equal dollar bets and measuring return to the long only). Therefore, the value of interest, a measure of the range of variation in the value of a spread bet or spread volatility, is directly computed from the spread return series, itself the numeric difference of the buy and sell return. (At this point of detail, one can begin to see how the considerations generalize beyond the pair setting to more general statistical arbitrages.)

Figure 6.1(c) shows the volatility traces, using a trailing 20-day window, of the two stocks ENL and RUK and of the spread ENL–RUK. (In all of the examples in this chapter, volatilities are computed under the conventional assumption of locally zero-mean return.) The spread volatility is, as foreshadowed, visually similar to the stock volatilities. Curiously, it is consistently greater than both the individual stock volatilities—more about that later.

Another example is shown in Figure 6.2, this time for the pair General Motors (GM) and Ford (F). Notice that the volatility of the spread is sometimes greater and sometimes less than the volatility of both stocks, and sometimes greater than one but less than the other.

These two examples expose many of the features of spread volatility that are important in understanding and exploiting spread relationships both for the simplest pairs, as illustrated, and more general cases including baskets of stocks. Figure 6.3 shows another example, this time of two unrelated stocks, Microsoft (MSFT) and EXXON (XON).

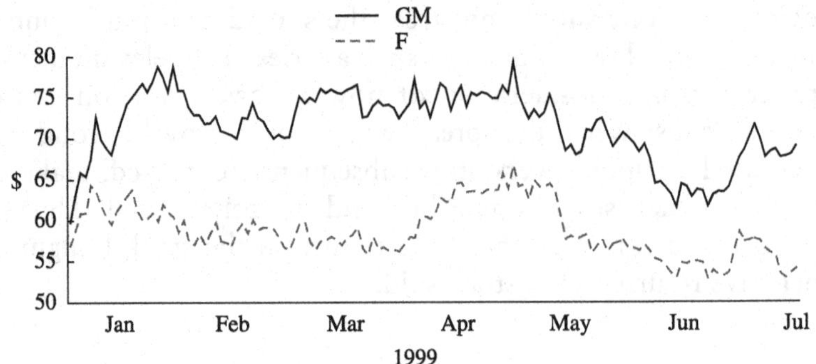

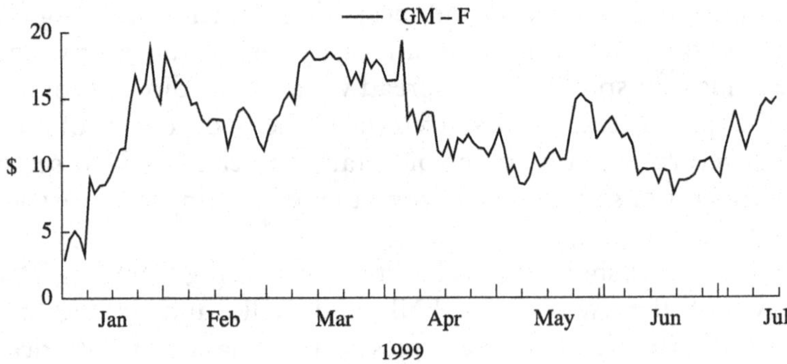

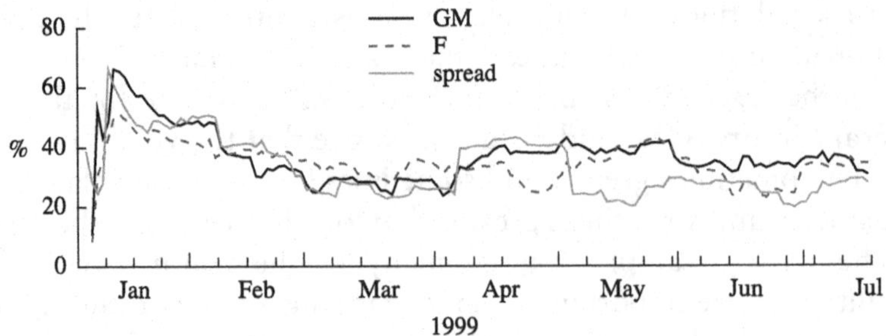

FIGURE 6.2 (a) Daily adjusted close prices of GM and F; (b) spread; (c) volatilities

6.2 THEORETICAL EXPLANATION

Relative price movement is functionally dependent on the price movements of individual stocks: What could be simpler than *price of A − price of B*? When looking at the variability of relative prices, the relationship is more complicated. The key relationship is that for the spread return already given:

$$\textit{spread return} = \textit{return on buy} - \textit{return on sell}$$

Writing A for the *return on buy*, B for the *return on sell*, and S for the *spread return*, the spread volatility is expressed as:

$$\begin{aligned}\sqrt{\mathrm{V}[S]} &= \sqrt{\mathrm{V}[A-B]}\\ &= \sqrt{\mathrm{V}[A]+\mathrm{V}[B]-2\mathrm{V}[A,B]}\end{aligned}$$

where $\mathrm{V}[\cdot]$ denotes (statistical or probabilistic) variance, and $\mathrm{V}[\cdot,\cdot]$ similarly denotes covariance. This expression immediately reveals how and why the spread volatility can be less than, greater than, or equal to the volatility of either constituent stock. The pertinent factor is the covariance of (the returns of) those two stocks, $\mathrm{V}[A, B]$.

If the two stocks A and B (abusing notation quite deliberately) are in fact the same stock, then the variances (the square of the volatility) are the same and, crucially, the covariance is also equal to the variance. Hence the volatility of the spread is zero: What else could it be since the spread itself is identically zero?

Now, what if the two stocks are unrelated? Statistically, this is equivalent to saying that the covariance is zero. Then the spread volatility reduces to:

$$\sqrt{\mathrm{V}[S]} = \sqrt{\mathrm{V}[A]+\mathrm{V}[B]}$$

That is, spread volatility is larger than both of the individual stock volatilities. If the individual stocks have similar volatility, $\mathrm{V}[A] \approx \mathrm{V}[B]$, then the inflation factor is about 40 percent:

$$\begin{aligned}\sqrt{\mathrm{V}[S]} &= \sqrt{\mathrm{V}[A]+\mathrm{V}[B]}\\ &\approx \sqrt{2\mathrm{V}[A]}\\ &= 1.414\sqrt{\mathrm{V}[A]}\end{aligned}$$

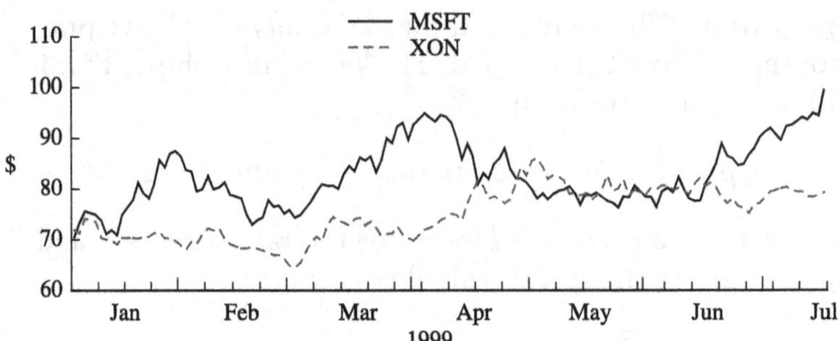

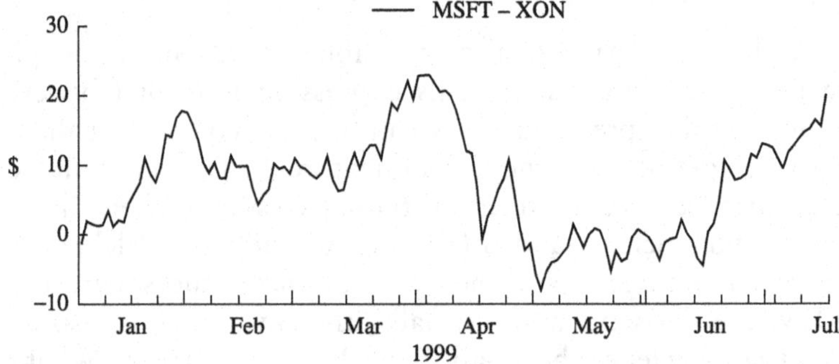

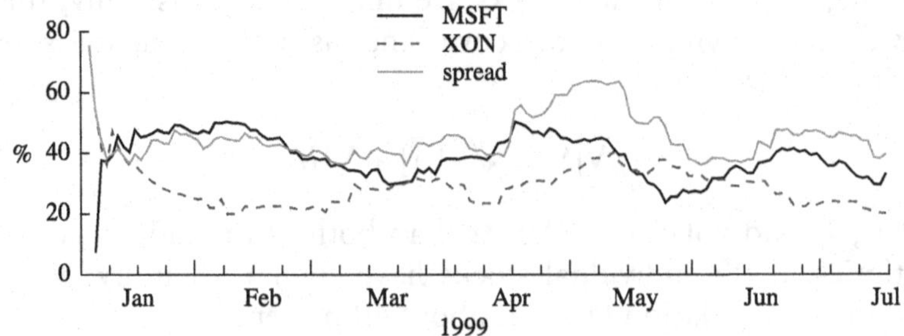

FIGURE 6.3 (a) Daily adjusted close prices of MSFT and XON; (b) spread; (c) volatilities

6.2.1 Theory versus Practice

The illustration in Figure 6.1 shows the spread volatility for two *related* stocks to be larger than both the individual stock volatilities. The theory advanced in the previous section says (1) spread volatility is zero for identical stocks, and (2) spread volatility is larger than both individual stocks for unrelated stocks. Ugh? Surely "related stocks" (such as ENL and RUK) are more like "identical stocks" than "unrelated stocks." So according to the theory, shouldn't the volatility of the ENL–RUK spread be small?

Now we must distinguish between statistical definitions of terms and English interpretations of the same terms. The two Elsevier stocks, ENL and RUK, are indeed related—essentially they are the same company. The historical traces of the price series show extraordinarily similar behavior as befits that. Over the long term, one is justified in stating that the prices are the same. However, the price traces on the daily time scale seldom move precisely in parallel; therefore the spread between the two does vary—seen in Figure 6.1(b)—and spread volatility is not zero. In fact, over the short term, the two price series show a negative relationship: In Figure 6.1(a) the two price traces proceed sinuously like two snakes entwined in a cartoon embrace, the one moving up when the other moves down and vice versa. Statistically, this means that the two series are negatively correlated, particularly on the short-term return scale which is pertinent to local volatility calculations.

Aha! Negative correlation (hence, negative covariance). Put that in the formula for spread volatility and immediately it is clear why the Elsevier stocks' spread volatility is greater than both the individual stock volatilities. Profit in the bank for pairs trading!

6.2.2 Finish the Theory

Return to the expression for spread volatility:

$$\sqrt{\mathrm{V}[S]} = \sqrt{\mathrm{V}[A] + \mathrm{V}[B] - 2\mathrm{V}[A,B]}$$

Write $\underline{\sigma}^2 = \min(\mathrm{V}[A], \mathrm{V}[B])$ and $\overline{\sigma}^2 = \max(\mathrm{V}[A], \mathrm{V}[B])$, then it is trivial to sandwich the spread volatility between multiples of the individual stock volatilities for uncorrelated stocks ($\mathrm{V}[A, B] = 0$):

$$\sqrt{2}\underline{\sigma} \leq \sqrt{\mathrm{V}[S]} \leq \sqrt{2}\overline{\sigma}$$

Two immediate observations have already been noted: For two similarly volatile stocks, the spread will exhibit 40 percent more volatility than the individual stocks; for two perfectly *positively* correlated stocks, the spread will exhibit no volatility because it is constant. That leaves one extreme case of relatedness: where A and B are perfectly *negatively* correlated, A = −B. Here the spread volatility is double the individual stock volatility:

$$\begin{aligned} \mathrm{V}[S] &= \mathrm{V}[A] + \mathrm{V}[B] - 2\mathrm{V}[A, B] \\ &= \mathrm{V}[A] + \mathrm{V}[-A] - 2\mathrm{V}[A, -A] \\ &= \mathrm{V}[A] + \mathrm{V}[A] + 2\mathrm{V}[A, A] \\ &= 4\mathrm{V}[A] \end{aligned}$$

Hence, $\sqrt{\mathrm{V}[S]} = 2\sqrt{\mathrm{V}[A]}$. For statistical arbitrage this is (almost) the grail of spread relationships.

6.2.3 Finish the Examples

What can be inferred about the GM–F and MSFT–XON examples with the benefit of the theory for the volatility of spreads? Given a description of the stock and spread volatility traces, one can point to periods of changing local correlation, positive to negative. Of course, one can observe correlation by direct calculation: See Figure 6.4. Average correlation in this first six months of 1999 is 0.58; maximum 20-day correlation is 0.86; minimum 20-day correlation is −0.15. Important to note here, for spread exploitation, are the dynamic changes in correlations and, hence, spread volatility and the range of variation.

From late April the GM–F spread volatility was less than both individual stock volatilities, as it was for most of March. In fact, from the spread trace in Figure 6.2(b) it is clear that for most of April and May, and again in June, the spread was practically constant in comparison to its value outside those periods. The spread volatility trace in Figure 6.2(c) shows a 50 percent hike in April and a similar drop in May. Clearly these are artifacts of the unusually large (negative) single day spread return on April 7 (see Figure 6.5) and the 20-day window used to compute an estimate of volatility—review the local correlation in Figure 6.4. Outlier down-weighting and smoothing are

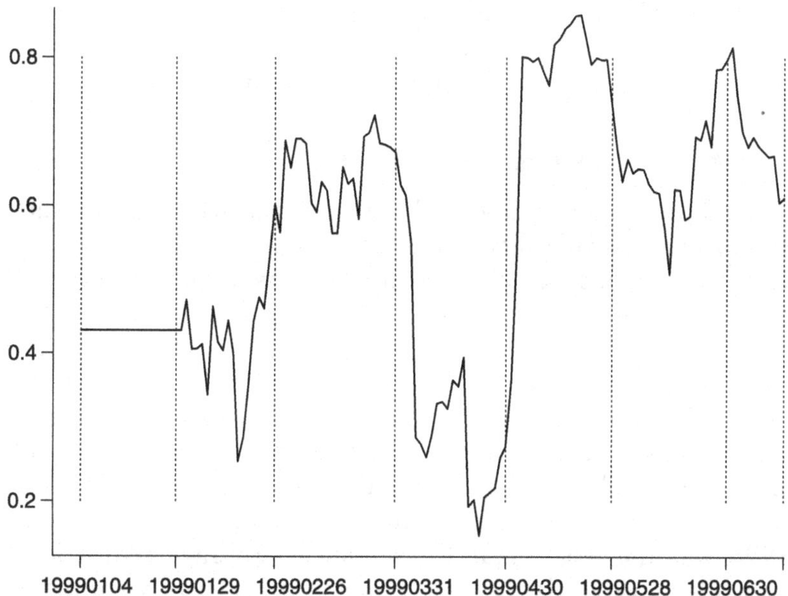

FIGURE 6.4 GM–F rolling 20-day correlation

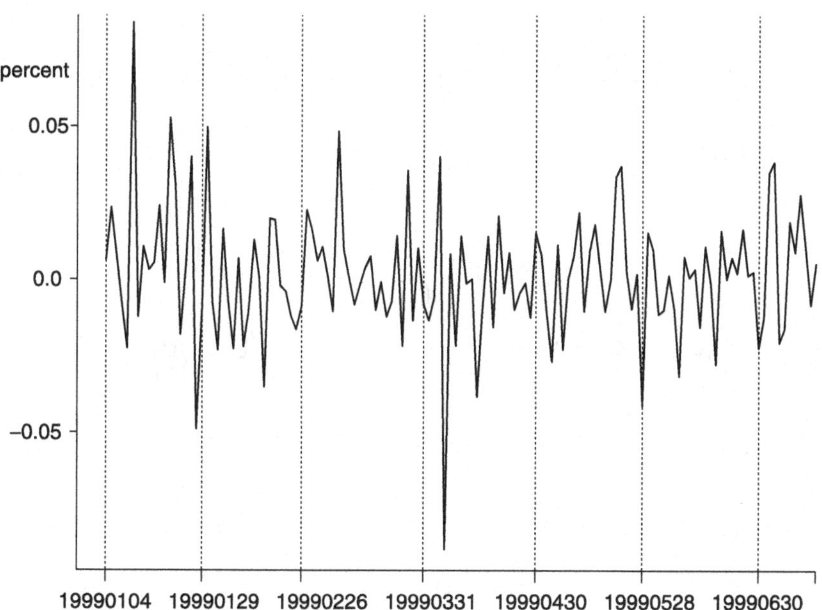

FIGURE 6.5 GM–F spread daily return

typical procedures used to reduce unrealistic jumps in such indirectly measured quantities as volatility (see Chapter 3). Figure 6.5 shows the spread return: return on GM minus return on F.

6.2.4 Primer on Measuring Spread Volatility

Let's begin by asking the question: Does statistical arbitrage generate higher returns when volatility is high or when it is low?

Absent any stock-specific events, higher interstock (spread) volatility should generate greater returns from a *well calibrated* model. Figure 6.6 shows the average local volatility (20-day moving window) for pairwise spreads for stocks in the S&P 500 index from 1995 through 2003. Two years of outstanding returns for statistical arbitrage were 2000 and 2001. Both were years of record high spread volatility; 2000 higher in spread volatility and statistical arbitrage return than 2001—nicely supporting the *ceteris paribus* answer. But 1999 was the worst year for statistical arbitrage return in a decade while spread volatility was equally high. There were many

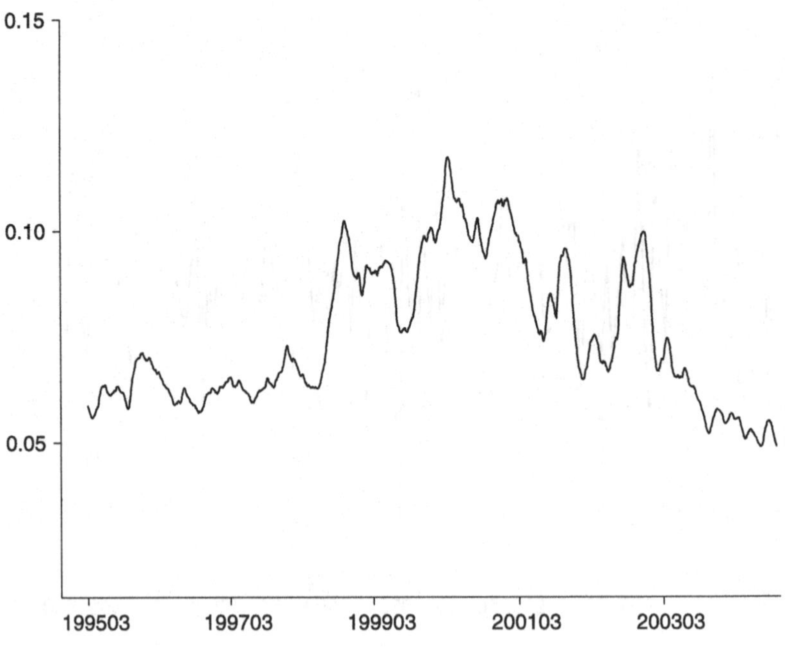

FIGURE 6.6 Average local standard deviation of spreads

stock-specific events, principally earnings related, with uniformly negative impact on return in 1999. So noticeable, widespread, and troubling were these events that the SEC eventually passed Regulation Fair Disclosure (Reg. FD) to outlaw the activities.

Using a local estimate of volatility, what picture is obtained from representative spread series? What can we infer from the spread volatility chart in Figure 6.6 using the sample local volatility reference patterns?

Figure 6.7 illustrates local volatility (using an equally weighted, 20-point window) for two sample spread series. The top panel, (a), shows the spread series, the center panel, (b), the local volatility estimates. There is nothing surprising here, the calculation being a measure of variation about a constant line segment of the curves in the top frame. Noteworthy is the observation that the average level of local volatility is similar for the two series.

What happens when a different measure of "local" is used? The bottom panel, Figure 6.7(c), illustrates the situation for a 60-point window: The striking feature now is the higher level of volatility indicated for the greater amplitude spread. (While we continue to couch the presentation in terms of a spread, the discussion applies equally to any time series.) Once again, there is no surprise here. The 60-point window captures almost a complete cycle of the greater amplitude series—the estimated volatility would be constant if precisely a full cycle was captured—and, hence, the local volatility estimate reflects the amplitude of the series. In the previous case, the shorter window was reflecting only part of the slower moving series variation. Which estimate of volatility reflects reversion return opportunity? Here the answer is easy.

Now consider what picture would emerge if an average over a set of such series were examined, each such series mixed with its own "noise" on both amplitude and frequency.

Properly cautioned, what can be inferred from Figure 6.6? Before attempting an answer, the archetypal example analyses clearly advise looking at local volatility estimates from a range of windows (or local weighting schemes)—it does seem advisable to concentrate on evidence from *shorter intervals* and focus on average levels of local volatility; mundane variation in the estimate may be little more than artifact. *May* be.

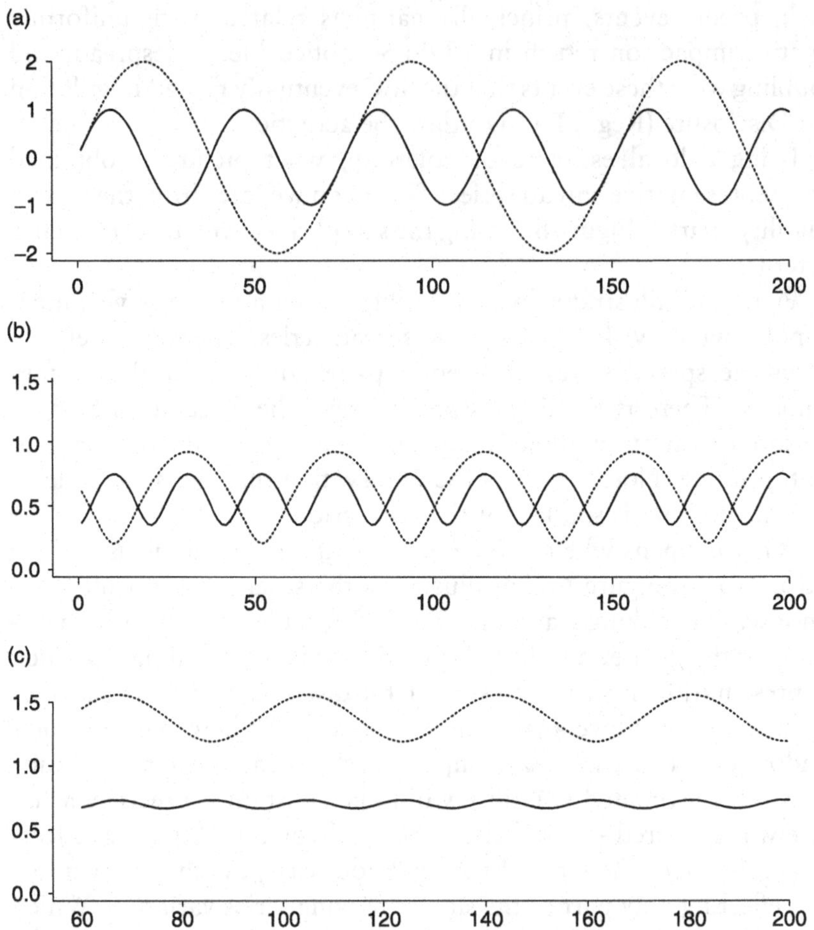

FIGURE 6.7 (a) Archetypal spread series; (b) local volatility estimate (20-point window, equal weights) of spread series; (c) local volatility estimate (60-point window) of spread series

Figure 6.8 reproduces the two example spread curves from Figure 6.7 and adds a third. The new series exhibits the same amplitude as the original high-amplitude series and the same frequency as the original higher frequency series. It therefore has the advantage of more frequent and higher value reversion opportunities. The center panel, (b), depicting local volatility estimates, indicates that the

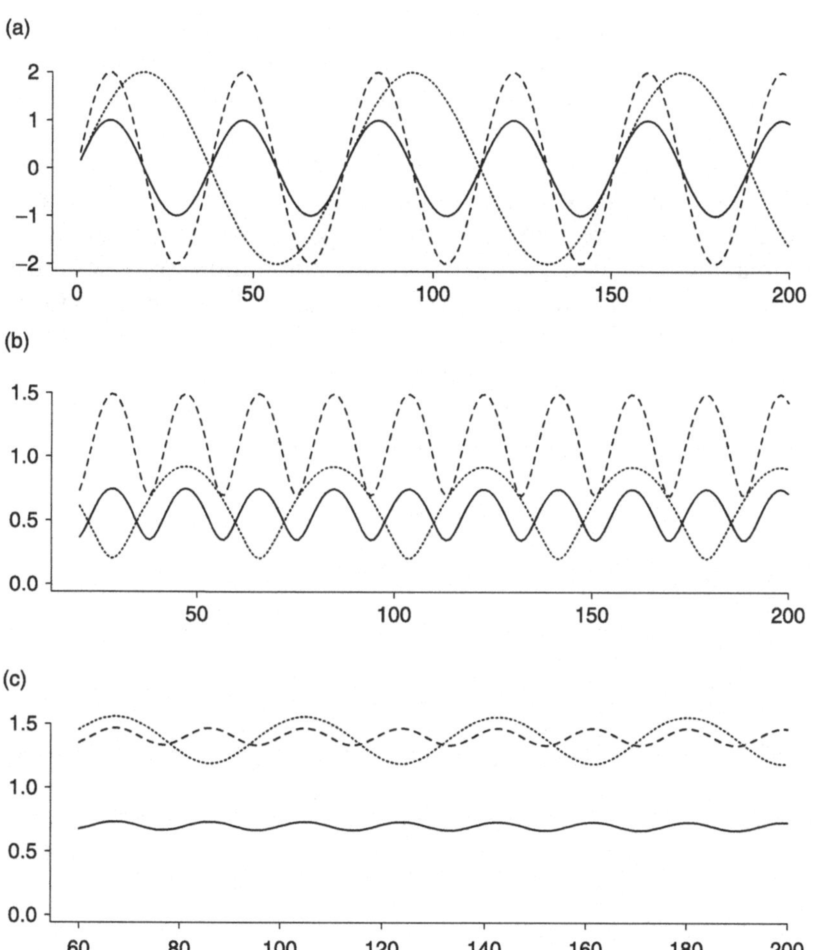

FIGURE 6.8 (a) Archetypal spread series; (b) local volatility estimate (20-point window, equal weights) of spread series; (c) local volatility estimate (60-point window) of spread series

average volatility of this third series is twice that of the original two, just as expected.

Now look at bottom panel, (c), which shows local volatility estimates from the longer window. Interesting? Once again, the analysis points to using a shorter, more local view when inferring reversion opportunity from average levels of spread volatility.

With these archetypal models, one can undertake an appropriate time-frequency analysis to precisely quantify the magnitude of reversionary moves. Real spread series are less obliging, beset with nonstationarity and "contaminating" noise.

The foregoing remarks are generally descriptive, characterizing how series variation is reflected in empirical summary statistics and indicating how the magnitude of potential gain from simple reversion plays may be estimated. Actual reversion exploitation strategies must be analyzed directly to make sensible inferences on expectations therefrom, whether in the idealized settings of noise-free sinusoidal series used here or in application to real spread histories.

Chapter 9 revisits interstock volatility in the context of the decline in statistical arbitrage performance since 2002.

CHAPTER 7

Quantifying Reversion Opportunities

Fortitudine vincimus—*By endurance we conquer.*
—Family motto of Sir E. H. Shackleton, polar explorer

7.1 INTRODUCTION

In this chapter, we extend the theoretical developments of the previous three chapters in the search for a deeper understanding of the properties of reversion in time series. There are more abstractions and more difficult mathematics here than elsewhere in the book, but in all cases, the theoretical development is subsequently grounded in application. Most of the discussion is framed in the language of price series, however, the developments generally apply to any time series. In particular, the analysis can readily be applied, sometimes with necessary revision of inference, to transformations of price series including returns, spreads, spread returns, factors, and so forth.

The question "What is reversion?" is addressed in the context of assumed probability models for stock price generation. The models are highly stylized and oversimplified, being entertained purely as a device for considering notions of reversion. There is no conceit that the models serve even as first pass approximations to the true, unknown, price generation mechanism. By determining the implications of definitions of reversion under the very restrictive assumptions of these simple models, it is hoped that a coherent view of what reversion is will emerge. The goal is to extract from such a view meaningful and quantifiable notions of reversion that may be used in the study of realistic price generation models. It is hoped that such understanding will provide new insight into statistical arbitrage, helping us to analyze and understand how and why statistical arbitrage works at a systems or mechanistic level, and from that build a valid market rationale for the driving forces of the exploitable opportunities. That may be a little ambitious; perhaps it is reaching to hope for more than indications of what kinds of processes to think about for such a rationale. The mechanics and the rationale are both critical to investigating and solving the problems that beset statistical arbitrage starting in 2004 and which continue to affect performance today: How do market structural changes impact strategy performance?

7.2 REVERSION IN A STATIONARY RANDOM PROCESS

We begin the study with consideration of the simplest stochastic system, a stationary random process. Prices are supposed to be generated independently each day from the same probability distribution, that distribution being characterized by unchanging parameters. We shall assume a continuous distribution. Price on day t will be denoted by P_t, lowercase being reserved for particular values (such as a realized price).

Some considerations that immediately suggest themselves are:

1. If P_t lies in the tail of the distribution, then it is likely that P_{t+1} will be closer to the center of the distribution than is P_t. In more formal terms: Suppose that $P_t >$ ninety-fifth percentile of the distribution. Then the odds that P_{t+1} will be smaller than P_t are 95 : 5 (19 : 1). A similar statement is obtained for the lower tail, of course.

   The 19 : 1 odds do not express quite the same idea as is expressed in the first sentence. Being "closer to the center than" is not the same as being "smaller than." Certainly the odds ratio quoted, and by implication the underlying scenario, are very interesting. For completeness, it is useful to examine the "closer to the center" scenario. The obvious notion of closer to the center is formally: the magnitude of the deviation from the center on the price scale. An alternative notion is to consider distance from the center in terms of percentiles of the underlying distribution of prices. The two notions are equivalent for distributions with symmetric density functions, but not otherwise.

2. If P_t is close to the center of the distribution, then it is likely that P_{t+1} will be further from the center than P_t.

After a little reflection, (ii) seems to offer infertile ground for a reversion study; but in a sequential context, values close to the center are useful flags for subsequent departure from the center and, hence, of future reversionary opportunities. Recall the popcorn process of Chapter 2 and the discussion of stochastic resonance in Chapter 3.

A generalization of the notion in (i) offers a potential starting point for the study: If $P_t >$ pth percentile of the distribution, then the odds that $P_{t+1} < P_t$ are $p : 100 - p$. Interest here is confined to those cases where the odds are better than even. Investors largely prefer strategies that exhibit more winning bets than losing bets, considering such relative frequency of outcomes a reflection of a stable process. The thought process is deficient because by itself the win–lose ratio imparts no information at all on the stability properties of a strategy other than the raw win–lose ratio itself. Essential information necessary for that judgment includes the description of the magnitudes of gains from winners and losses from losers. A strategy that loses 80 percent of the time but that never exhibits individual losses exceeding 0.1 percent and whose winners always gain 1 percent is a stable and profitable system. Judgments about easily labeled but complicated notions such as "stability" need careful specification of personal preferences. Often these are not made explicit and are therefore readily miscommunicated because individuals' preferences are quite variable.

For $P_t >$ median, the odds that $P_{t+1} < P_t$ are greater than one; similarly, for $P_t <$ median, the odds that $P_{t+1} > P_t$ are also greater than one. The assumption of continuity is critical here, and a reasonable approximation for price series notwithstanding the discrete reality thereof. You may want to revisit Chapter 4 for a rehearsal of the difficulties discrete distributions can pose.

Two questions are immediately apparent:

1. Is the odds result exploitable in trading?
   - With artificial data following the assumed stationary process.

   - With stock data using locally defined moments (to approximate conditional stationarity).
2. How should reversion be defined in this context?
   - Reversion *to* the center requires modification of the foregoing odds to something like 75 percent → 50 percent and 25 percent → 50 percent.
   - Reversion *in the direction of* the center—so that overshoot is allowed and the odds exhibited are pertinent.

Whichever of (1) or (2) is appropriate (which in the context of this chapter must be interpreted as "useful for the analysis and interpretation of reversion in price series"), how should reversion be characterized? As the proportion of cases exhibiting the required directional movement (a distribution free quantity)? As the expected (average) price movement in the qualifying reversionary cases (which requires integration over an assumed price distribution and is not distribution-free)?

Both aspects are important for a trading system. In a conventional trading strategy, betting on direction and magnitude of price movements, total profits are controlled by the expected amount of reversion. If the model is reasonable, the greater the expected price movement, the greater the expected profit. The volatility of profits in such a system is determined in part by the proportion of winning to losing moves. For the same unconditional price distributions, the higher the ratio of winners to losers, the lower the profit variance thus spreading profit over more winning bets and fewer losing bets. Stop loss rules and bet sizing significantly impact outcome characteristics of a strategy, making more general absolute statements unwise.

It is worth noting the following observation. Suppose we pick only those trades that are profitable round trip. Daily profit variation will still, typically, be substantial. Experiments with real data using a popcorn process model show that the proportion of winning days can be as low as 52 percent for a strategy with 75 percent winning bets and a Sharpe ratio over 2.

Reversion defined as any movement from today's price in the direction of the center of the price distribution includes overshoot cases. The scenario characterizes as reversionary movement a price greater than the median that moves to *any* lower price—including to any price lower than the median, the so-called overshoot. Similarly, movement from a price below the median that moves to *any* higher price is reversionary.

7.2.1 Frequency of Reversionary Moves

For any price greater than the median price, $P_t = p_t > m$:

$$\Pr[P_{t+1} < p_t] = F_P(p_t)$$

where $F_P(\cdot)$ denotes the distribution function of the probability distribution from which prices are assumed to be generated. (This result is a direct consequence of the independence assumption.) An overall measure of the occurrence of reversion in this situation is then:

$$\int_m^\infty F_P(p_t) f_P(p_t) \mathrm{d}p_t = \frac{3}{8}$$

where $f_p(.)$ is the density function of the price distribution. Therefore, also considering prices less than the median, $P_t < m$, we might say that *reversion occurs 75 percent of the time*. This is the result proved and discussed at length in Chapter 4.

Previously it was noted that the proportion of reversionary moves is a useful characterization of a price series. The 75 percent result states that underlying distributional form makes no difference to the proportion of reversionary moves. Therefore, low volatility stocks will exhibit the same proportion of opportunities for a system exploiting reversion as high volatility stocks. Furthermore, stocks that are more prone to comparatively large moves (outliers, in statistical parlance) will also exhibit the same proportion of reversionary opportunities as stocks that are not so prone. The practical significance of this result is that the proportion of reversionary moves is not a function of the distribution of the underlying randomness. Thus, when structure is added to the model for price generation, there are no complications arising from particular distribution assumptions. Moreover, when analyzing real price series, observed differences in the proportion of reversionary moves unambiguously indicate differences in temporal structure other than in the random component.

7.2.2 Amount of Reversion

Following the preceding discussion of how often reversion is exhibited, some possible measures of the size of expected reversion from a price greater than the median are:

1. $\mathrm{E}[P_t - P_{t+1} | P_t > P_{t+1}]$ Expected amount of reversion, given that reversion occurs.
2. $\mathrm{E}[P_t - P_{t+1} | P_t > m]$ Average amount of reversion.
3. $\mathrm{E}[P_t - P_{t+1} | P_t > P_{t+1}] \mathrm{Pr}[P_{t+1} < P_t]$ Overall expected amount of reversion.

Remarks: $P_t > m$ is an underlying condition. The difference between cases 1 and 2 is that case 2 includes the 25 percent of cases where $P_t > m$ but reversion does not occur, $P_{t+1} > P_t$, while case 1 does not. Case 1 includes *only* reversionary moves.

If case 1 is taken to define the total amount of "pure" reversion in the system, then case 2 may be considered as the "revealed" reversion in the system. With this terminology, it is possible to envisage a system in which the revealed reversion is zero or negative while the pure reversion is always positive (except in uninteresting, degenerate cases).

Moves from a price less than the median are characterized analogously.

Pure Reversion Expected pure reversion is defined as:

$$\mathrm{E}[P_t - P_{t+1} | P_{t+1} < P_t, P_t > m] \times \frac{1}{2}$$
$$+ \mathrm{E}[P_{t+1} - P_t | P_{t+1} > P_t, P_t < m] \times \frac{1}{2}$$

The two pieces correspond to (a) downward moves from a price above the median and (b) upward moves from a price below the median. It is important to keep the two parts separate because the expected values of each are generally different; only for symmetric distributions are they equal. Consider the first term only. From Figure 7.1, the cases of interest define the conditional distribution represented by the shaded region. For any particular price $P_t = p_t > m$, the expected amount

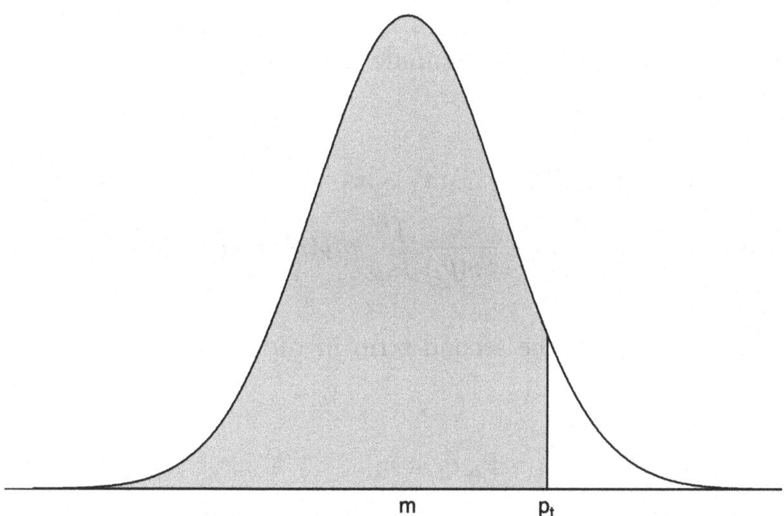

FIGURE 7.1 Generic price distribution

of reversion is just P_t minus the expected value of the conditional distribution:

$$p_t - \mathrm{E}_{P_{t+1}|P_{t+1}<p_t}[P_{t+1}] = p_t - \int_{-\infty}^{p_t} p_{t+1} f_{P_{t+1}|P_{t+1}<p_t}(p_{t+1})\mathrm{d}p_{t+1}$$

The density of the conditional distribution of P_{t+1}, given that $P_{t+1} < P_t$, is just the rescaled unconditional density (from the independence assumption), the scale factor being one minus the probability of the subset of the original domain excluded by the conditioning. Expected reversion is therefore:

$$p_t - \frac{1}{F_P(p_t)} \int_{-\infty}^{p_t} p f_P(p)\mathrm{d}p$$

Now we are interested in the expected value of this quantity averaged over all those possible values $P_t = p_t > m$:

$$\begin{aligned}
&\mathrm{E}_{P_t>m}\left[P_t - \mathrm{E}_{P_{t+1}|P_{t+1}<P_t}[P_{t+1}]\right] \\
&= \int_m^{\infty} \left(p_t - \frac{1}{F_P(p_t)} \int_{-\infty}^{p_t} p f_P(p)\mathrm{d}p\right) f_{P_t|P_t>m}(p_t)\mathrm{d}p_t
\end{aligned}$$

Substituting for $f_{P_t|P_t>m}(p_t) = f_P(p_t)/(1 - F_P(m)) = 2f_P(p_t)$ for $P_t > m$, then the overall expected amount of pure (one-day) reversion when $P_t > m$ is:

$$\begin{aligned}
&\mathrm{E}[P_t - P_{t+1}|P_{t+1} < P_t, P_t > m] \\
&= 2\int_m^{\infty}\left(p_t - \frac{1}{F_P(p_t)}\int_{-\infty}^{p_t} pf_P(p)\mathrm{d}p\right) f_P(p_t)\mathrm{d}p_t
\end{aligned}$$

A similar analysis for the second term in the original expectation yields:

$$\begin{aligned}
&\mathrm{E}[P_{t+1} - P_t|P_{t+1} > P_t, P_t < m] \\
&= 2\int_{-\infty}^{m}\left(\frac{1}{1 - F_P(p_t)}\int_{p_t}^{\infty} pf_P(p)\mathrm{d}p - p_t\right) f_P(p_t)\mathrm{d}p_t
\end{aligned}$$

Adding (one half times) these two results gives the expected pure reversion as:

$$\begin{aligned}
&\int_m^{\infty}\left(p_t - \frac{1}{F_P(p_t)}\int_{-\infty}^{p_t} pf_P(p)\mathrm{d}p\right) f_P(p_t)\mathrm{d}p_t \\
&+ \int_{-\infty}^{m}\left(\frac{1}{1 - F_P(p_t)}\int_{p_t}^{\infty} pf_P(p)\mathrm{d}p - p_t\right) f_P(p_t)\mathrm{d}p_t
\end{aligned}$$

Some simplification of this expression would be nice. Example 1, which follows, shows simplification possible for the normal distribution, the double integral here reducing to a single integral; however, even there, a closed form solution remains elusive. The symmetry is suggestive. Would the result simplify if the cut-off is taken as the mean rather than the median? Certainly an assumption of a symmetric density leads to cancellation of the two direct terms in P_t; in fact, the two parts of the sum are equal. Perhaps Fubini's rule, reversing the order of integration, can usefully be applied? We do know that the result is positive! A closed-form theoretical result remains unknown at this time, but computation of any specific example is straightforward (see the examples that follow).

Revealed Reversion Expected revealed reversion is defined as:

$$E[P_t - P_{t+1}|P_t > m] \times \frac{1}{2} + E[P_{t+1} - P_t|P_t < m] \times \frac{1}{2}$$

Consider the first term of the expression:

$$\begin{aligned} E[P_t - P_{t+1}|P_t > m] &= E[P_t|P_t > m] - E[P_{t+1}|P_t > m] \\ &= E[P_t|P_t > m] - E[P_{t+1}] \qquad \text{by independence} \\ &= E[P_t|P_t > m] - \mu \end{aligned}$$

where μ denotes the mean of the price distribution. Similarly, the second term of the expression reduces to $E[P_{t+1} - P_t|P_t < m] = \mu - E[P_t|P_t > m]$. It is worth noting that both terms have the same value, which follows from:

$$\begin{aligned} \mu = E[P_t] &= \int_{-\infty}^{\infty} p f_P(p) dp \\ &= \int_{-\infty}^{m} p f_P(p) dp + \int_{m}^{\infty} p f_P(p) dp \\ &= \frac{1}{2} E[P_t|P_t < m] + \frac{1}{2} E[P_t|P_t > m] \end{aligned}$$

whereupon:

$$\begin{aligned} E[P_t|P_t > m] - \mu &= E[P_t|P_t > m] - \frac{1}{2} E[P_t|P_t < m] - \frac{1}{2} E[P_t|P_t > m] \\ &= \frac{1}{2} E[P_t|P_t > m] - \frac{1}{2} E[P_t|P_t < m] \\ &= \frac{1}{2} E[P_t|P_t > m] + \frac{1}{2} E[P_t|P_t < m] - E[P_t|P_t < m] \\ &= \mu - E[P_t|P_t < m] \end{aligned}$$

Total expected revealed reversion may therefore be expressed equivalently as:

$$\begin{aligned} \textit{total expected revealed reversion} &= 0.5 \times (E[P_t|P_t > m] - E[P_t|P_t < m]) \\ &= E[P_t|P_t > m] - \mu \\ &= \mu - E[P_t|P_t < m] \end{aligned}$$

which (for continuous distributions) is positive except in uninteresting, degenerate cases.

Note 1: This result provides a lower bound for case 1 since the latter excludes those outcomes, included here, for which the actual reversion is negative, namely $\{P_{t+1} : P_{t+1} > P_t, P_t > m\}$ and $\{P_{t+1} : P_{t+1} < P_t, P_t < m\}$.

Note 2: A desirable property for the reversion measure to have is invariance to location shift. The amount of reversion, in price units, should not change if every price is increased by $1. It is easy to see that the expression for expected revealed reversion is location invariant: Moving the distribution along the scale changes the mean and median in the same amount. For the pure reversion result, it is not very easy to see the invariance from the equation. However, consideration of Figure 7.1 fills that gap.

Some Specific Examples

Example 1 Prices are normally distributed. If X is normally distributed with mean μ and variance σ^2, then the conditional distribution of X such that $X < \mu$ is the half normal distribution. Total expected revealed reversion is 0.8σ. (The mean of the truncated normal distribution is given in Johnson and Kotz, Volume 1, p. 81; for the half normal distribution the result is $E[X)] = 2\sigma/\sqrt{2\pi}$.) Thus, the greater the dispersion of the underlying price distribution, the greater the expected revealed reversion: a result that is nicely in tune with intuition and desire.

From a random sample of size, 1,000 from the standard normal distribution, the sample value is 0.8, which is beguilingly close to the theoretical value. Figure 7.2 shows the sample distribution.

The pure reversion general result can be reduced somewhat for the normal distribution. First, as already remarked, the terms in P_t cancel because the density is symmetric, leaving:

$$-\int_m^\infty \left(\frac{1}{F(p_t)}\int_{-\infty}^{p_t} pf(p)\mathrm{d}p\right) f(p_t)\mathrm{d}p_t$$

$$+\int_{-\infty}^m \left(\frac{1}{1-F(p_t)}\int_{p_t}^{\infty} pf(p)\mathrm{d}p\right) f(p_t)\mathrm{d}p_t$$

(The subscript on f and F has been dropped since it is not necessary to distinguish different conditional and unconditional distributions

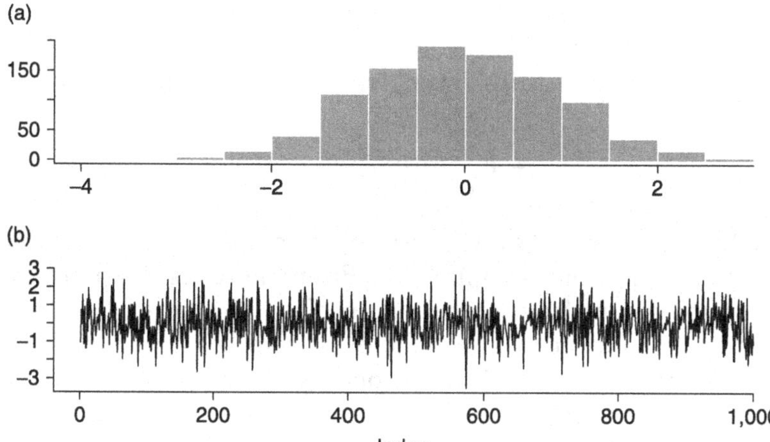

FIGURE 7.2 Random "price" series: (a) sample distribution, (b) sequential ordering

here: Only the unconditional price distribution is used.) Johnson and Kotz give results for moments of truncated normal distributions. In particular, the expected values of singly truncated normals required here are:

$$\int_{-\infty}^{p_t} pf(p)\mathrm{d}p = -\frac{f(p_t)}{F(p_t)}$$

and

$$\int_{p_t}^{\infty} pf(p)\mathrm{d}p = \frac{f(p_t)}{1 - F(p_t)}$$

Therefore, expected pure reversion is:

$$-\int_{m}^{\infty} \frac{1}{F(p)} \frac{-f(p)}{F(p)} f(p)\mathrm{d}p + \int_{-\infty}^{m} \frac{1}{1 - F(p)} \frac{f(p)}{1 - F(p)} f(p)\mathrm{d}p$$

$$= \int_{m}^{\infty} \left(\frac{f(p)}{F(p)}\right)^2 \mathrm{d}p + \int_{-\infty}^{m} \left(\frac{f(p)}{1 - F(p)}\right)^2 \mathrm{d}p$$

For a symmetric density, inspection shows that the two terms in the sum are equal. Algebraically, noting that $f(m - \epsilon) = f(m + \epsilon)$ and $F(m - \epsilon) = 1 - F(m + \epsilon)$, then a simple change of variable, $q = -p$, gives the result immediately. The quantity $(1 - F(x))/f(x)$ is known as Mills' ratio. Therefore, expected pure reversion for the normal independent identically distributed (*iid*) model is twice the integral of the inverse squared Mills' ratio with integration over half the real line:

$$2\int_{-\infty}^{m} M(p)^{-2}\mathrm{d}p$$

Panel (b) of Figure 7.2 shows the random sample referred to at the beginning of the section as a time series. From this series: The number of downward moves from above the median, $\{x_t : x_t > 0 \cap x_{t+1} < x_t\}$, is 351; the number of upward moves from below the median, $\{x_t : x_t < 0 \cap x_{t+1} > x_t\}$, is 388; the proportion of reversionary moves is $100 * (351 + 388)/999 = 74\%$ (the denominator is reduced by one because of the need to work with pairs (x_t, x_{t+1}) and, of course, there is no value to pair with $x_{1,000}$).

Revealed reversion for this time series Figure 7.3 shows the distribution of one-day "price" differences for (a) moves from above the median, $\{x_t - x_{t+1} : x_t > 0\}$, and (b) moves from below the median, $\{x_{t+1} - x_t : x_t < 0\}$. Not surprisingly, for such a large sample the two look remarkably similar (in fact, the sums, or sample estimates of expected values, agree to four significant figures); the total of these moves is $794/999 = 0.795$, which is very close to the theoretical expected value of 0.8. Some discrepancy is expected because treating the random sample as a time series imposes constraints on the components of the sets of pairs of values comprising the moves.

Pure reversion for this time series Figure 7.4 shows the distribution of one-day "price" differences for (a) moves from above the median *in a downward direction only*, $\{x_t - x_{t+1} : x_t > 0 \cap x_t > x_{t+1}\}$, and (b) moves from below the median *in an upward direction only*, $\{x_{t+1} - x_t : x_t < 0 \cap x_t < x_{t+1}\}$. These histograms are simply truncated versions of those in Figure 7.3, with 0 being the truncation point. Total pure reversion is $957/999 = 0.95$.

This example data is further discussed in Example 5.

Example 2 Prices are distributed according to the Student t distribution on 5 degrees of freedom. A Monte Carlo experiment yielded the expected revealed reversion to be 0.95 (for the unit scale t_5 distribution). Notice that this value is larger than the corresponding

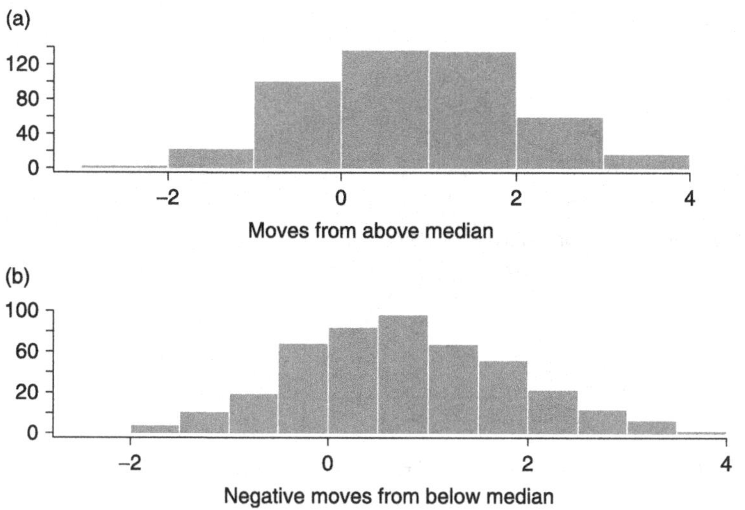

FIGURE 7.3 Random "price" series: distribution of one-day moves

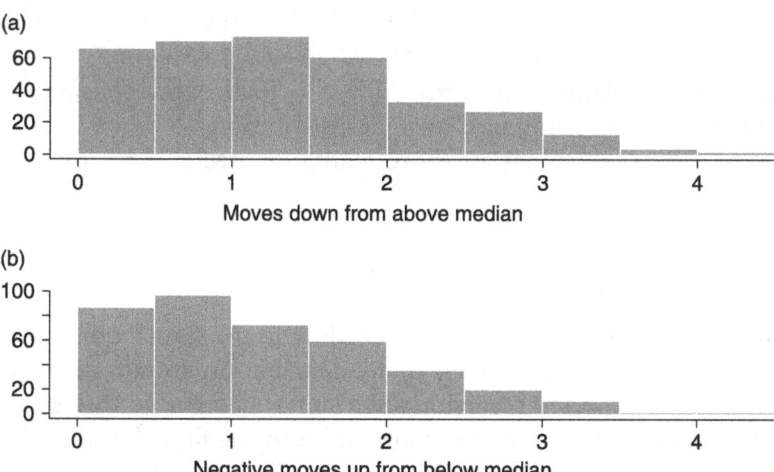

FIGURE 7.4 Random "price" series: distribution of one-day moves

value for the unit scale (standard) normal distribution (0.8). The increase is a consequence of the pinching of the t density in the center and stretching in the tails in comparison with the normal: The heavier tails mean that more realizations occur at considerable

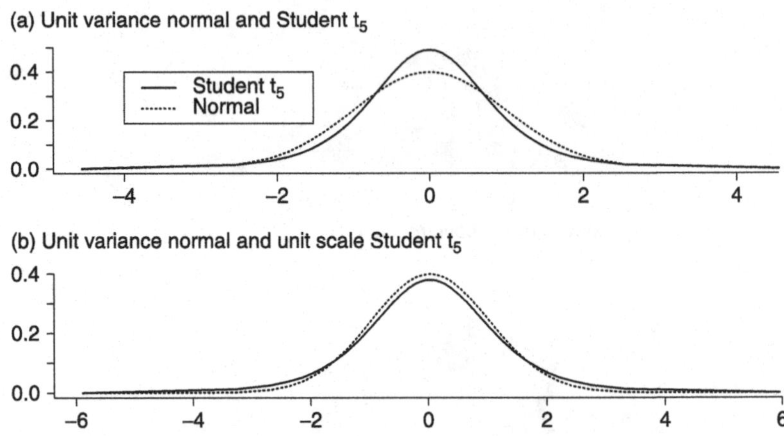

FIGURE 7.5 Comparison of normal and Student t distributions

distance from the center. Recall that the variance of the t distribution is the scale multiplied by $dof/(dof-2)$ where dof denotes the degrees of freedom; the unit scale t_5 distribution has variance $\frac{5}{3}$. Thus, the t_5 distribution with scale $\frac{3}{5}$ has unit variance and exhibits revealed reversion (in this sample) of $0.95 \times \sqrt{3/5} = 0.74$, which is *smaller* than the value for the standard normal distribution.

These comparisons may be more readily appreciated by looking at Figure 7.5.

Pure reversion See the remarks in Example 1.

Example 3 Prices are distributed according to the Cauchy distribution. Since the moments of the Cauchy distribution are not defined, the measures of expected reversion are also not defined, so this is not a fruitful example in the present context. Imposing finite limits—the truncated Cauchy distribution—is an interesting intellectual exercise, one that is best pursued under the aegis of an investigation of the t distribution, since the Cauchy is the t on one degree of freedom.

Example 4 An empirical experiment. Daily closing prices (adjusted for dividends and stock splits) for stock AA for the period 1987–1990 are shown in Figure 7.6. Obviously, the daily prices are not independent, nor could they reasonably be assumed to be drawn from the

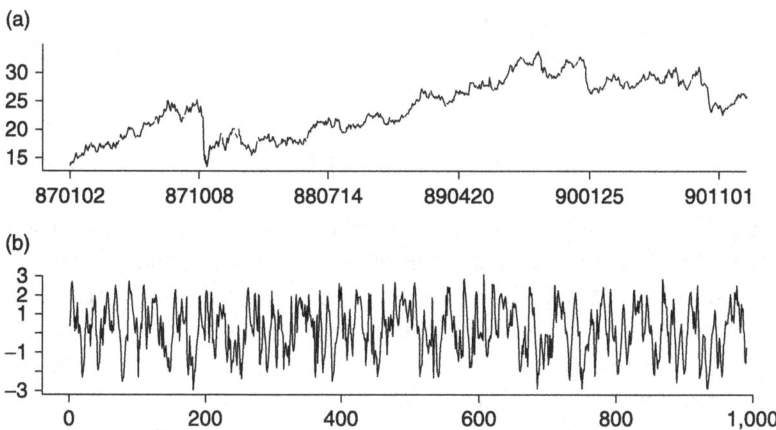

FIGURE 7.6 Daily close price for stock AA (adjusted for dividends and stock splits): (a) actual, (b) standardized for local median and standard deviation

same distribution. These violations can be mitigated somewhat by locally adjusting the price series for location and spread. Even so, it is not expected that the 75 percent reversion result will be exhibited: Serial correlation structure in the data is not addressed for one thing. The point of interest is just how much reversion actually is exhibited according to the measures suggested. (An unfinished task is apportionment of the departure of empirical results from the theoretical results to the several assumption violations.)

The daily price series is converted to a standardized series by subtracting a local estimate of location (mean or median) and dividing by a similarly local estimate of standard deviation. The local estimates are based on an exponentially weighted moving average of recent past data: In this way an operational online procedure is mimicked (see Chapter 3). Figure 7.6 shows the standardized price series using an effective window length of 10 business days; the location estimate is the median. Compare this with Figure 7.8, referred to later, which shows the price series adjusted for location only.

For the location-adjusted series, *not* standardized for variance, the proportion of reversionary moves is 58 percent, considerably less than the theoretical 75 percent. Note that 0 is used as the median of the adjusted series. By construction, the median should be close to zero; more significantly, the procedure retains an operational facility by this choice. A few more experiments, with alternative weighting schemes using effective window lengths up to 20 business days and using the local mean in place of the local median for location estimate, yield similar results: the proportion of reversionary moves being in the range 55 to 65 percent. The results clearly suggest that one or more of the theorem assumptions are not satisfied by the local location adjusted series.

Figure 7.7 shows the distribution of price changes (close–previous close) for those days predicted to be reverting downward and (previous close–close) for those days predicted to be reverting upward. The price changes are calculated from the raw price series (not median adjusted) since those are the prices that will determine the outcome of a betting strategy. Figure 7.7 therefore shows the distribution of raw profit and loss (P&L) from implementing a betting strategy based on stock price movement relative to local average price. Panel (a) shows the distribution of trade P&L for forecast downward reversions from a price above the local median, panel (b) shows the distribution of trade P&L for forecast upward reversions from a price below the local median. Clearly neither direction is, on average, profitable. In sum, the profit is \$−31.95 on 962 trades of the \$15–30 stock (there are 28 days on which the local median adjusted price is 0, and $2 \times k = 20$ days are dropped from the beginning of the series for initialization of local median and standard deviation thereof).

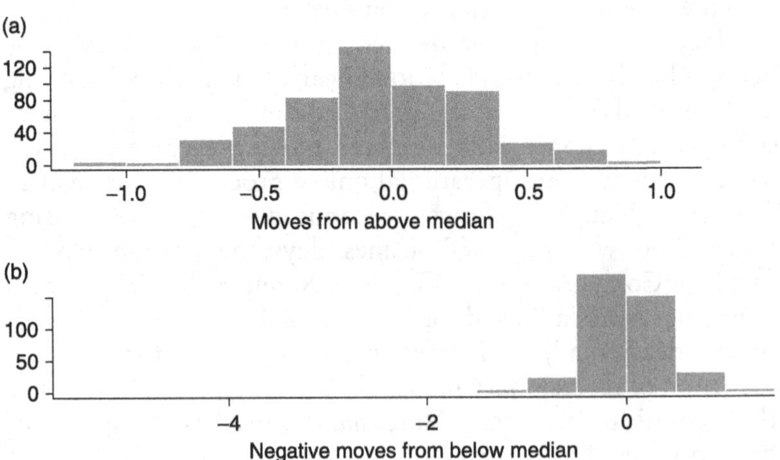

FIGURE 7.7 Stock AA one-day moves around local median: (a) from above the median and (b) from below the median

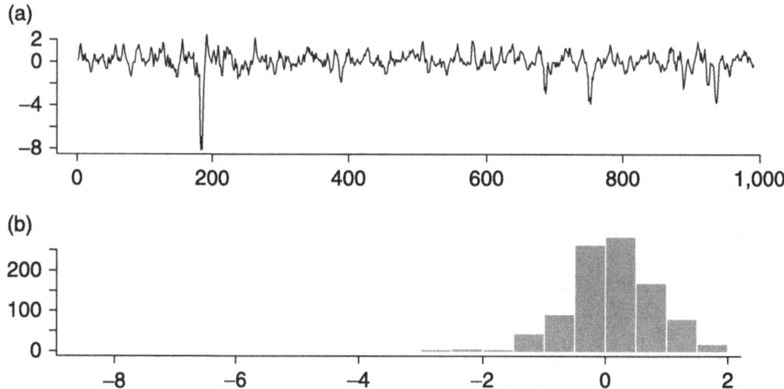

FIGURE 7.8 Stock AA local median adjusted close price: (a) time series and (b) histogram

Figure 7.8 shows the distribution of price minus local median. Taking this as the distribution in Section 7.2 (with the median therein becoming zero), the total revealed reversion is \$597.79/990 = \$0.60 per day. The actual result reported in the preceding paragraph, \$−31.95, shows the extent to which assumption violations (with unmodeled or poorly modeled structure) impact expected revealed reversion in this example.

This "missing structure" impact is perhaps more readily appreciated from a pure analysis of the median adjusted series. The revealed reversion from this series (in contrast to the already reported reversion from the raw price series given signals from the adjusted series) is \$70.17. This means that less than one-eighth of the revealed reversion from the independent, identically distributed model is recoverable from the local location adjusted data series. (Outlier removal would be a pertinent exercise in a study of an actual trading system. In an operational context, the large outlier situations would be masked by risk control procedures.) Example 5 attempts to uncover how much reversion might be lost through local location adjustment.

Figure 7.9 shows the sample autocorrelation and partial autocorrelation estimates: Evidently there is strong 1- and 2-day serial correlation structure in the locally adjusted series. Undoubtedly that

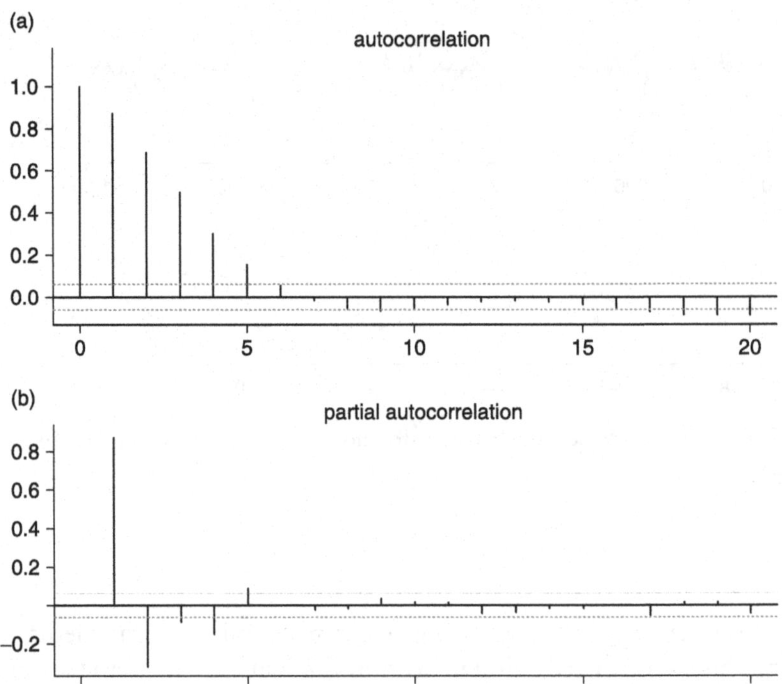

FIGURE 7.9 Stock AA: (a) autocorrelations and (b) partial autocorrelations of local median adjusted close price

accounts for part (most?) of the deviation of actual revealed reversion from the theoretically expected value under the assumption of independence.

Pure reversion For the record, total pure reversion in the median adjusted price series (from the actual time series, since we do not yet have a closed-form result to apply to the price-move distributions, although we do know that it must exceed the sample estimate of the theoretical value of revealed reversion or \$597.79) is \$191.44 on 562 days. Pure reversion from the raw price series (using signals from the adjusted series) is just 60 percent of this at \$117.74.

Example 5 This is an extended analysis of the data used in Example 1. In Example 4, the operational procedure of local median

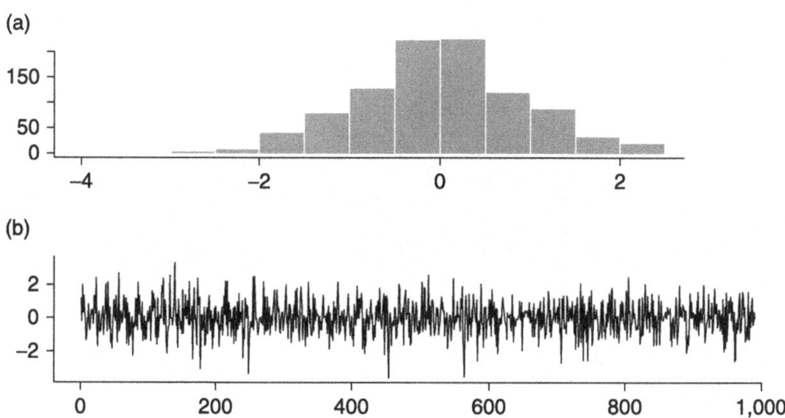

FIGURE 7.10 Random price series, local median analogs of Figure 7.2: (a) histogram and (b) time series

adjustment was introduced as a pragmatic way of applying the 75 percent result to real, but nonstationary, data series. It is of interest to understand the implications of the operational procedure for a series that is actually stationary. Such knowledge will help determine the relative impact of empirical adjustment for location against other assumption violations such as serial correlation.

Figures 7.10 to 7.12 are local median adjusted versions of Figures 7.2 to 7.4. (The local median is calculated from a window of the preceding 10 data points.) The summary statistics, with values from the original analysis in Example 1 in parentheses, are: 77 percent (74 percent) of reversionary moves; total pure reversion = 900(952); total revealed reversion = 753(794).

The analysis depicted in Figures 7.13 to 7.15 is perhaps more pertinent. The median adjusted series determines when a point is above or below the local median but, in contrast to the case reported in the preceding paragraph, the raw, unadjusted series is used to calculate the amount of reversion. This is the situation that would be obtained in practice. Signals may be generated by whatever model one chooses, but actual market prices determine the trading result. Interestingly, pure reversion increases to 906—but this value is still

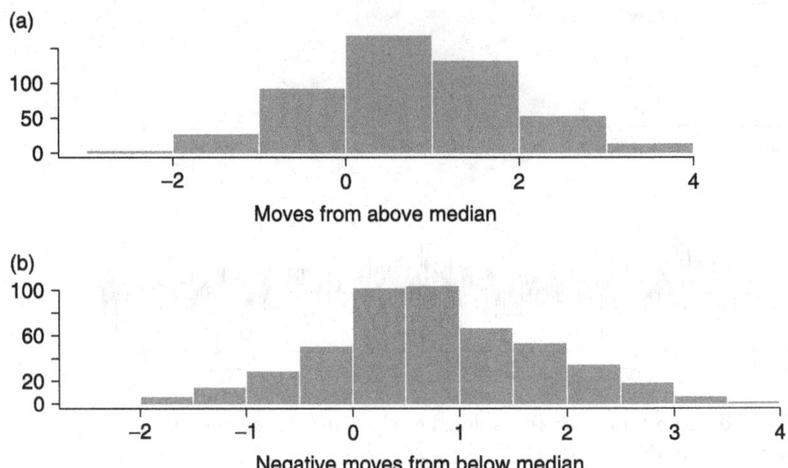

FIGURE 7.11 Random price series, local median analogs of Figure 7.3: (a) moves from the median and above the median (b) moves from below the median

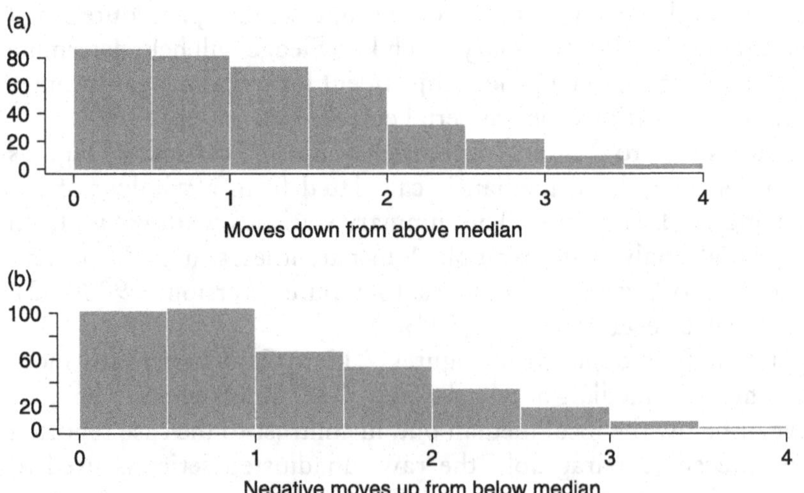

FIGURE 7.12 Random price series, local median analogs of Figure 7.4: (a) moves from above the median and (b) moves from below the medium

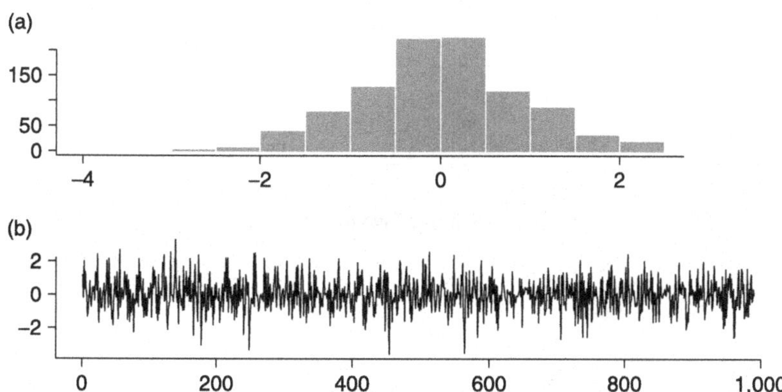

FIGURE 7.13 Random price series, signals from local median adjusted series with reversion from raw series: (a) histogram and (b) time series

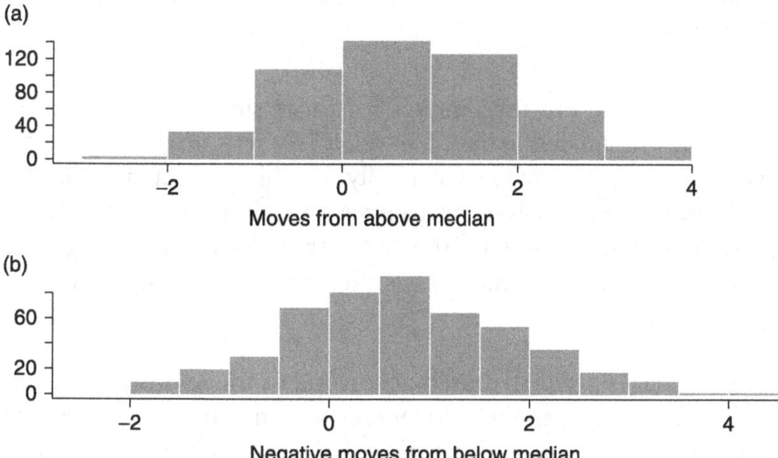

FIGURE 7.14 Random price series, signals from local median adjusted series with reversion from raw series: (a) moves from above median and (b) moves from below median

well below the unadjusted series result of 952. Revealed reversion decreases to 733.

Figure 7.15 is interesting. Notice that there are *negative* "moves down from above the median" which is logically impossible! This reflects the fact that the signals are calculated from the local median

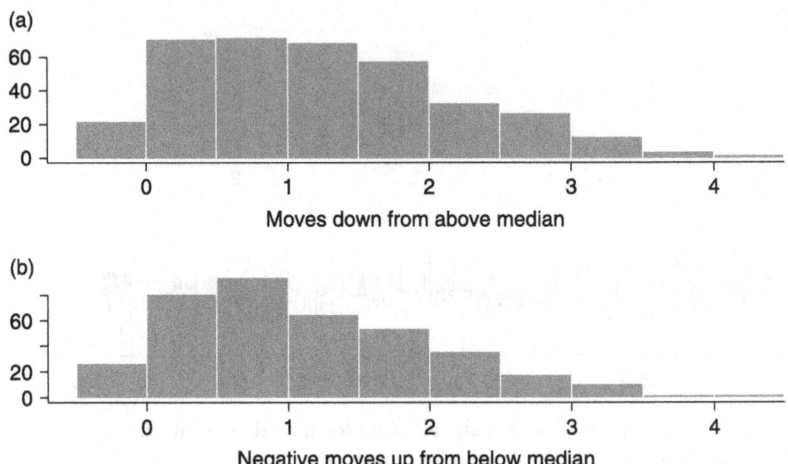

FIGURE 7.15 Random price series, signals from local median adjusted series with reversion from raw series: (a) moves from above the median and (b) moves from below the median

adjusted price series, but then moves for those signals are calculated from the raw, unadjusted series. The relatively few small magnitude, negative moves is expected. Curiously in this particular sample, despite the negative contributions to pure reversion, the total actually increases; that is the result of the more-than-offsetting changes in magnitude of the positive moves (in the raw series compared with the standardized series).

Example 6 Prices are distributed according to the lognormal distribution. If log X is normally distributed with mean μ and variance σ^2, then X has the lognormal distribution with mean $\mu_X = exp(\mu + 1/2\sigma^2)$, variance $\sigma_X^2 = exp(2\mu + \sigma^2)(exp(\sigma^2) - 1)$, and median $\exp(\mu)$. Using results for the truncated lognormal distribution from Johnson and Kotz, Volume 1, p.129, total expected revealed reversion is:

$$\exp(\mu + \sigma^2/2)[1 - \Phi(-\sigma)]$$

where $\Phi(.)$ denotes the cumulative standard normal distribution function. Details are given in Appendix 7.1 at the end of this chapter. Figure 7.16 shows the histogram and time series representation of a random sample of 1,000 draws from the lognormal distribution

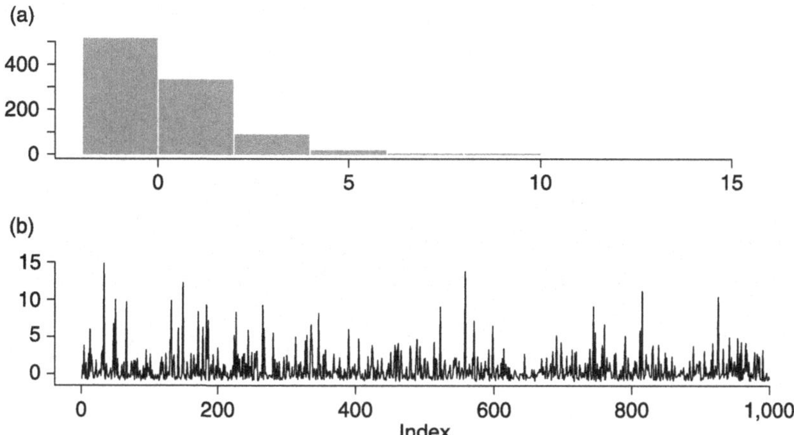

FIGURE 7.16 Random sample from lognormal distribution (with median subtracted): (a) histogram and (b) time series

with $\mu = 0$ and $\sigma = 1$. (The median, 1, is subtracted to center the distribution on 0.) The sample value of expected revealed reversion is 1.08 (theoretical value is 1.126). Treating the sample as a time series in the manner of Example 5, the sample estimate of pure reversion is 1.346.

7.2.3 Movements from Quantiles Other Than the Median

The analysis thus far has concentrated on all moves conditional on today's price being above or below the median. Figure 7.17 shows that, for the normal and Student t distributions, the median is the sensible focal point. For example, if we consider the subset of moves when price exceeds the sixtieth percentile (and, by symmetry here, does not exceed the fortieth percentile), the expected price change from today to tomorrow is less than the expected value when considering the larger subset of moves that is obtained when price exceeds the median.

It is expected that this result will *not* remain true when serially correlated series are examined. Trading strategies must also consider transaction costs, which are not included in the analysis here.

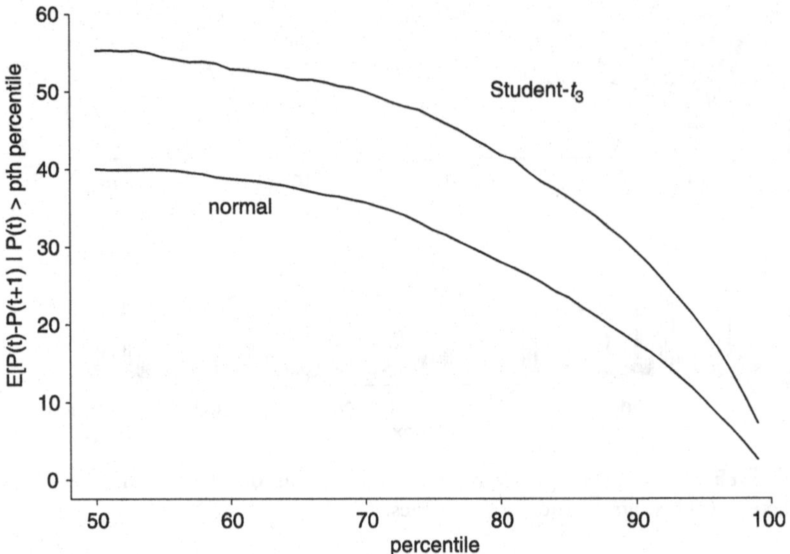

FIGURE 7.17 Expected pure reversion by conditioning percentile of price distribution

7.3 NONSTATIONARY PROCESSES: INHOMOGENEOUS VARIANCE

The very stringent assumptions of the strictly stationary, independent, identically distributed (*iid*) process examined in Section 7.2 are relaxed in this section. Here we generalize the measures of pure and revealed reversion to the inhomogeneous variance analog of the *iid* process. See Chapter 4 for generalization of the "75 percent theorem."

Prices are supposed to be generated independently each day from a distribution in a given family of distributions. The family of distributions is fixed but the particular member from which price is generated on a given day is uncertain. Members of the family are distinguished only by the variance. Properties of the realized variance sequence now determine what can be said about the price series.

7.3.1 Sequentially Structured Variances

Conditional on known, but different, variances, each day a normalized price series could be constructed, then the results of Section

7.2 would apply directly to that normalized series. Retrospectively, it is possible to compare a normalized price series with theoretical expectations (much as we did in Section 7.3 where normalization was not required); it is also possible to calculate actual pure and revealed reversion, of course. However, it is not clear that one can say anything prospectively and therefore construct a trading rule to exploit expected reversions.

One possibility is to look at the range of variances exhibited and any systematic patterns in the day-to-day values. In the extreme case where it is possible to predict very well the variance tomorrow, a suitable modification of the 75 percent rule and calculation of pure and revealed reversion is possible. Such calculations would be useful providing that the same (or, in practice, very similar) circumstances (variance today versus variance tomorrow) occur often enough to apply expected values from probability distributions. This may be a realistic hope as variance clusters frequently characterize financial series: See Chapter 3 for modeling possibilities.

As a particularly simple special case of perfect variance prediction, suppose that the only variance inhomogeneity in a price series is that variances on Fridays are always twice the value obtained for the rest of the week. In this case, there is no need to expend effort on discovering modified rules or expected reversion values: The 75 percent rule and calculations of expected pure and revealed reversion apply for the price series with Fridays omitted. Recall that we are still assuming independence day to day so that selective deletion from a price series history does not affect the validity of results. In practice, serial correlation precludes such a simple solution. Moreover, why give up Friday trading if, with a little work, a more general solution may be derived?

7.3.2 Sequentially Unstructured Variances

This is the case analyzed in detail in Chapter 4, Section 3. For the normal-inverse Gamma example explored there (see Figure 4.2) the expected reversion calculations yield the following. Actual revealed reversion is $206.81/999 = 0.21$ per day; pure reversion is $260.00/999 = 0.26$ per day. Notice that the ratio of pure to revealed, $0.26/0.21 = 1.24$, is larger than for the normal example (Example 1) in Section 7.3, $0.94/0.8 = 1.18$.

7.4 SERIAL CORRELATION

In Section 7.3, the analysis of stock AA (Example 4) showed that the price series, actually the local median adjusted series, exhibited strong first-order autocorrelation, and weaker but still notable second-order autocorrelation. We commented that the presence of that serial correlation was probably largely responsible for the discrepancy between the theoretical results on expected reversion for *iid* series (75 percent) and the actual amount calculated for the (median adjusted) series (58 percent). The theoretical result is extended for serial correlation in Section 4 of Chapter 4. We end the chapter here using the example from Chapter 4.

Example 7 One thousand terms were generated from a first-order autoregressive model with serial correlation parameter $r = 0.71$ and normally distributed random terms (see Example 1 in Chapter 4). Figure 4.4 shows the time plot and the sample marginal distribution.

The proportion of reversionary moves exhibited by the series is 62 percent; total revealed reversion is 315 and total pure reversion is 601. Ignoring serial correlation and using the sample marginal

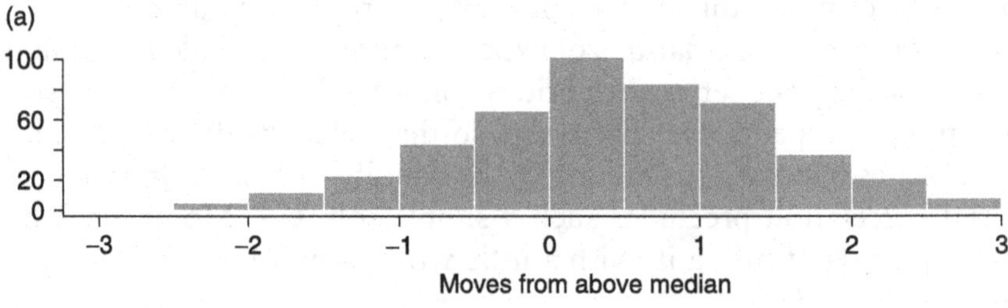

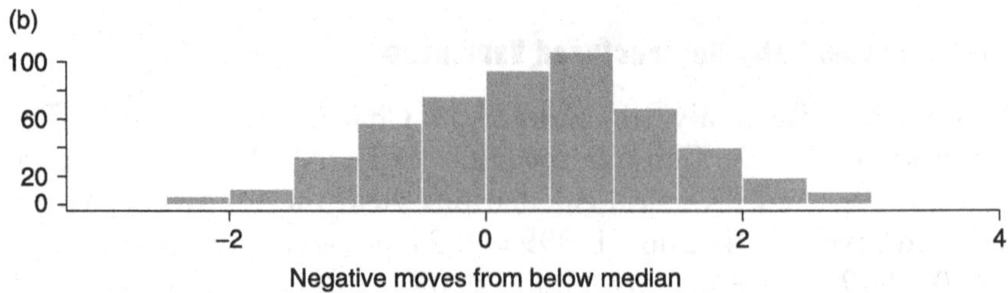

FIGURE 7.18 Analysis of autocorrelated series, Example 7: (a) moves from above the median and (b) moves from below the median

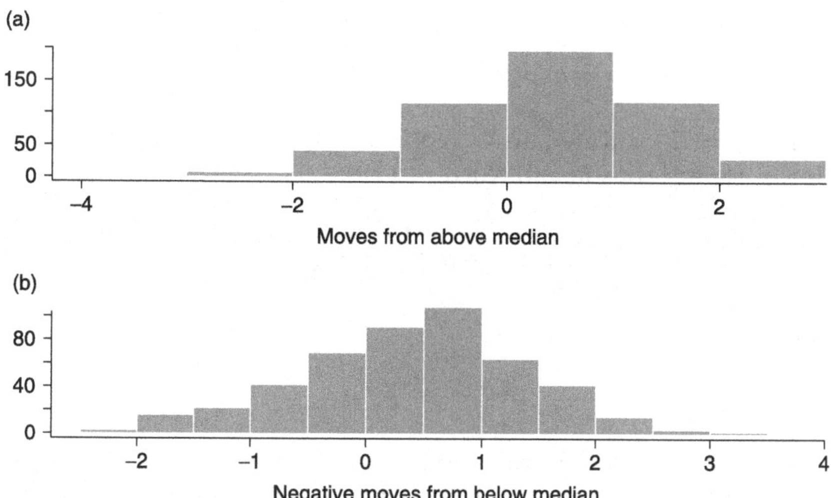

FIGURE 7.19 Analysis of autocorrelated model, local median adjusted: (a) moves from above the median and (b) moves from below the median

distribution to calculate the result in Section 7.2.2, the theoretical revealed reversion is 592—almost double the actual value. Figure 7.18 illustrates aspects of the analysis.

The local median adjusted series (window length of 10) is shown in Figure 4.4; aspects of the reversion assessment are illustrated in Figure 7.19. A slightly greater proportion of reversionary moves is exhibited by the adjusted series, 65 percent (compared with 62 percent previously noted for the raw series). Total revealed reversion is 342 (compared to 315 in the unadjusted series); total pure reversion is 605 (compared to 601).

APPENDIX 7.1: DETAILS OF THE LOGNORMAL CASE IN EXAMPLE 6

$$Y = \log X \sim N[\mu_Y, \sigma_Y^2]$$

Set:

$$\mathrm{E}[X] = \mu_X = \exp(\mu_Y + \sigma_Y^2/2)$$

$$V[X] = \sigma_X^2 = \exp(2\mu_Y + \sigma_Y^2)(\exp(\sigma_Y^2) - 1)$$
$$median = \exp(\mu_Y)$$

Define $Z = X$ truncated at X_0 (equivalently, Y is truncated at $Y_0 = logX_0$. Then (Johnson and Kotz, p. 129):

$$E[Z] = \mu_Z = \mu_X \frac{1 - \Phi(U_0 - \sigma_Y)}{1 - \Phi(U_0)}$$

where:

$$U_0 = \frac{\log X_0 - \mu_Y}{\sigma_Y}$$

Total expected revealed reversion can be written as $E[X|X > m] - E[X]$. Now, $E[X|X > m] = E[Z]$ with $X_0 = m = exp(\mu_Y)$. In this case, U_0 reduces to 0 and:

$$\mu_Z = \mu_X \frac{1 - \Phi(-\sigma_Y)}{1 - \Phi(0)} = 2\mu_X[1 - \Phi(-\sigma_Y)]$$

Therefore, total expected revealed reversion is:

$$\begin{aligned} 2\mu_X[1 - \Phi(-\sigma_Y)] - \mu_X &= \mu_X[1 - \Phi(-\sigma_Y)] \\ &= \exp(\mu_Y + \sigma_Y^2/2)[1 - \Phi(-\sigma_Y)] \end{aligned}$$

Special Case $\mu_Y = 0, \sigma_Y = 1$. Then $\mu_X = \sqrt{e}, \sigma_X^2 = e(e-1), X_0 =$ median $= e^0 = 1$. Now, $U_0 = logX_0 = 0$, so that:

$$\mu_Z = \sqrt{e}\frac{1 - \Phi(-1)}{1 - \Phi(0)} = 2\sqrt{e}[1 - \Phi(-1)]$$

From standard statistical tables (see references in Johnson, Kotz, and Balakrishnan), $\Phi(-1) = 0.15865$ so the mean of the median truncated lognormal distribution (with $\mu_Y = 0, \sigma = 1$) is 2.774.

CHAPTER 8

Nobel Difficulties

Chance favors the prepared mind.

—Louis Pasteur

8.1 INTRODUCTION

In this chapter, we examine scenarios that create negative results for statistical arbitrage plays. When operating an investment strategy, and notwithstanding risk filters and stop loss rules, surprises should be expected to occur with some frequency. The first demonstration examines a single pair that exhibits textbook reversionary behavior until a fundamental development, a takeover announcement, creates a breakpoint. Next we discuss the twofold impact of an international economic development, the credit crisis of 1998: introducing a new risk factor into the equity market—temporary price discrimination as a function of credit rating on corporate debt—and turning a profitable year (to May) into a negative year (to August). Next we consider how large redemptions from funds such as hedge, mutual, and pension, create temporary disruptions to stock price dynamics with deleterious effects on statistical arbitrage performance. Next we relate the story of Regulation FD. Finally, in all this discussion of performance trauma we revisit the theme of Chapter 5, clearing up misunderstandings, specifically on the matter of correlation of manager performance in negative return periods.

8.2 EVENT RISK

Figure 8.1 shows the price histories (daily close price, adjusted for stock splits and dividends) for stocks Federal Home Loan Mortgage Corporation (FRE) and Sunamerica, Inc. (SAI) from January 2, 1996 to March 31, 1998. The two price traces track each other closely with a strong upward trend, the spread between the two repeatedly widening and closing.

The analysis and demonstration that follows focuses on pair spread trading, but the salient points on structural shifts are relevant more widely to statistical arbitrage models. A factor model forecasting individual stock movements (in groups) is similarly vulnerable to the described motions, but the mechanics are more involved and explanation requires greater subtlety of detail. We will keep the analysis simple and remind the reader once more that the basic points are applicable more generally to statistical arbitrage models.

Daily returns for the two stocks do not correlate particularly strongly; the correlation is 0.4. However, looking at returns between events, the correlation is much higher at 0.7. Event correlation

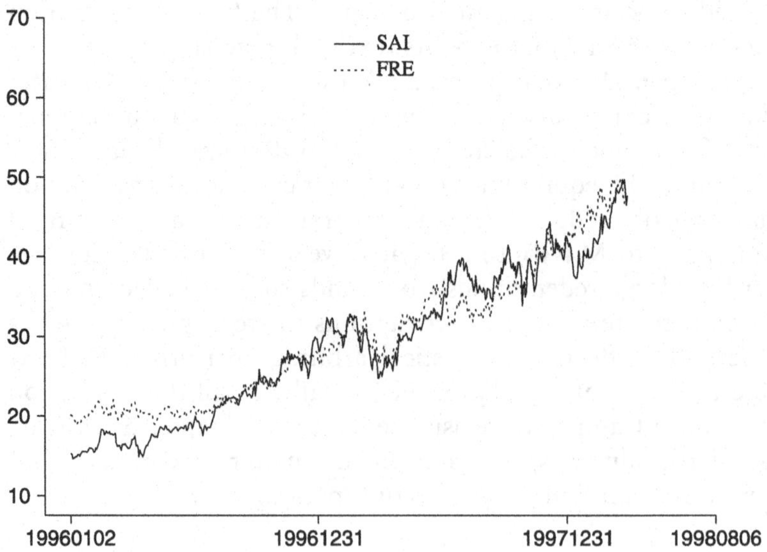

FIGURE 8.1 Adjusted price histories for FRE and SAI

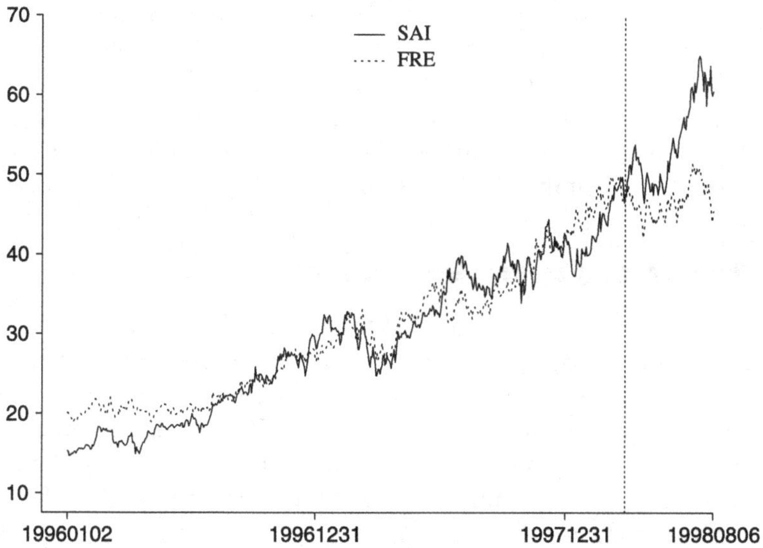

FIGURE 8.2 Adjusted price histories for FRE and SAI to August 1998

indicates what might be expected to trade well in groups within prescribed risk tolerances (see Chapter 2).

Visually and statistically, it looks as though the pair [FRE, SAI] will trade profitably in a simple spread reversion model. Simulation of a basic popcorn process model (see Chapter 2) demonstrates that was indeed the case.

Figure 8.2 shows the adjusted price series for FRE and SAI extended through the second quarter of 1998 to August 6. Interesting? The spread widened considerably to more than double the recent historical maximum. As already noted, the size of the spread does not give rise to losing trades, the process of spread widening does. When a spread begins a persistent period of growth, trades tend to be unwound (a) after a longer than usual period and (b) when the local mean spread is substantially different from where it was when the trade was entered, causing a loss. (This analysis assumes autopilot application of the forecast model without intervention schemes. Including monitors and related early exit [stop loss] rules would attenuate losses but complicate the description without changing the basic message.)

Using an exponentially weighted moving average model for an assumed popcorn process with a constrained trend component (see Chapter 3) trades in [FRE, SAI] entered in late February lasted through late April, and trades entered in early June lasted through early July. Both occasions returned substantial losses. (There was a fast turnaround, profitable trade in late July.)

8.2.1 Will Narrowing Spreads Guarantee Profits?

Sadly, there are no guarantees of profitable scenarios. However, one beneficial asymmetry of decreasing volatility compared to increasing volatility is that in the former case the model's lagged view works to advantage when the local mean is changing. (Recall that when the local mean is not changing, changes in volatility are not a problem, though there may be opportunity costs, in the form of missed profits, if volatility forecasts are slow to adapt.)

When the spread local mean is persistently moving in one direction in excess of the model's limited pace of adaptation, the strategy loses because the trade exit point (zero expected return) is bad relative to the trade entry. Contemporaneous entry and exit points exhibit the proper relationship; the problem arises from changes over time as the spread develops differently from the forecast made at the time of the trade entry. If the prediction of local volatility is greater than actual volatility, then current trade entries will be conservative (fewer actual trades, and each with higher expected return). When the trend continues to the disadvantage of the model, this conservatism reduces losses, quite the reverse of what happens when volatility is increasing and the model is underestimating it.

Cutting off losing trades when spread relationships change is ideal. However, implementing this requires another prediction: a prediction of the change. Typically the best we can do is identify change soon after it happens, and characterize it soon after that. Even this is challenging. See Chapter 3 and the referenced discussion in Pole et al., 1994.

Looking at current FRE–SAI trades from the perspective of August 1998 we ask: Must we maintain a persistently losing position? Obviously not; a position can be unwound at a manager's discretion. But when should the pair be readmitted to the candidate trade universe? The historically tight coupling of FRE and SAI seems to be breaking down; if the breakdown applies to the underlying common structure of the stock return series, then the stocks will cease to satisfy pair selection criteria and trading will cease. If the common structure remains, with spreads oscillating about a new level or returning to the recent historical level, the stocks will continue to be selected and will again be traded profitably once the disturbance is over.

On Thursday August 20, 1998 the takeover of SAI by AIG was announced. SAI closed on Wednesday at $64 $\frac{3}{8}$ the all stock deal valued SAI at a 25 percent premium (before the market opening). One might seriously wonder at the nature of the buying pressure behind the run-up in price of SAI before the takeover.

8.3 RISE OF A NEW RISK FACTOR

The international credit crisis of the summer of 1998 was an interesting time for the practice of statistical arbitrage. Performance problems began in June and, for many, accelerated through July and August. During this time, it became starkly apparent for the first time that perceptions about the credit quality of a company had a direct impact on investor confidence in near-term company valuation. As sentiment became negative and stocks were marked down across the market, the magnitude of price decline was strongly related to the credit rating of outstanding corporate debt. Companies with lower rating had their stock price decimated by a factor of three to one more than companies with a higher rating. Such dramatic discriminatory action had not previously been described; certainly there was no prior episode in the history of statistical arbitrage.

There are many hypotheses, fewer now entertained than was the case at the time, about the nature of the linkages between credit and equity markets in 1998, and why the price movements were so dramatic. Without doubt, the compounding factor of the demise of the hedge fund Long-Term Capital Management and the unprecedented salvage operation forced by the Federal Reserve upon unenthusiastic investment banks heightened prevalent fears of systemic failure of the U.S. financial system. At the naive end of the range of hypotheses is the true, but not by itself sufficient, notion that the Fed's actions simply amplified normal panic reactions to a major economic failing. An important factor was the speed with which information, speculation, gossip, and twaddle was disseminated and the breadth of popular coverage from twenty-four hour "news" television channels to the Internet. The large number of day traders and active individual participants drawn into the market during the 1990s by the attraction of the bull run, and trading made easy by technological developments, provided a receptive audience for the broadcast noise and a fertile environment in which to germinate and breed fear and panic. There was much poor decision making borne of instantaneous judgments, couched as "analysis" though often little more than the rambling of the moment to fill immediate desire for sound bites, speedily implemented by the facilitating technology. Good for volatility and studies of lemming-like behavior in cognitively higher order species, bad for blood pressure and ulcers.

When the impact of the Russian default began to be experienced on the U.S. stock markets, concern grew that companies would be progressively squeezed in the market for credit and this concern led to stock price markdowns. As the crisis continued and stock prices declined, the prices of companies with lower credit rating declined faster and cumulatively by more than prices of companies with higher credit rating, existentially proving the prevalent fear, rational or not, that tightening credit markets (the link from the Russian default to this outcome internationally or specifically in the United States not convincingly, coherently made) would make raising finance more expensive. And what could be more logical than that poorer rated companies would have to pay more? The apparent logic for discriminatory stock price realignment looks unassailable on the surface. Since that is as far as much "analysis" went (and goes) the consequences were those of self fulfilling prophecy. Was there ever a likelihood of U.S. interest rates being raised as a result of the Russian default?

Corporate debt rating became a significant discriminatory factor in U.S. equities in the summer of 1998. Any portfolio constructed absent attention to this factor was likely to be exposed to valuation loss directly as the lower rated stocks' prices declined more than proportionately compared with higher rated stocks. Whether constructed as a collection of matched pairs from a vanilla pair trading strategy or from a sophisticated factor-based return prediction model makes no difference at the outset. Losses are inevitable.

As the discriminatory stock price patterns developed, discriminatory results distinguished types of statistical arbitrage strategy, though manager action in the face of persistent and cumulatively large losses complicates assessment: the mix of model versus manager (to the extent that is meaningful) being impossible to identify. Simulation studies devoid of manager contamination indicate that factor models exhibited more resilience[1] than pure spread-based models, and quicker resumption of positive returns.

With recognition of a new risk factor, what should be done? Factor models, when the factor decomposition is recomputed using return history from the period affected, will naturally incorporate "debt rating," so direct action is not necessary. Inaction does beg a few questions though: What should be done during the evolving episode once the factor has been identified (or posited at least)? Is the best one can do simply to wait for a new window of data from which to estimate stock exposures to the factor (and meanwhile take a performance wallop to the chin)? Answer to the latter is obvious but, beyond a simple "No," sensible prescriptions are more demanding to compose. General specification of the foremost requirement is direct: Eliminate exposure to the posited factor from the portfolio. Precise action to accomplish that is a tad more difficult—what *are* the exposures? In the haste necessitated by the strong emotional push and business need to staunch losses, luck played its role.

---

[1]From where does this greater resilience derive? A partial answer can be constructed by contrasting a basic pair strategy with a basic factor model strategy (which models are precisely the source of the simulation results on which the evidential commentary is made). The pair portfolio consists of bets on pairs of stocks that are matched on standard fundamental measures including industry classification, capitalization, and price–earnings ratio. A first-order DLM forecast model is assumed for the spread (using the log price ratio series), with an approximate GARCH-like variance law. Bets are made when the spread deviates from its forecast by 15 percent or more (annualized). All signaled bets are taken and held until the model generates an exit signal; bets are not rebalanced; no stop loss rule is applied. The factor model is constructed as described in Chapter 3, with the optimization targeting annualized 15 percent for comparison with the pair model. Positions are rebalanced daily according to forecasts and factor exposures.

There is some evidence that credit rating was correlated with a combination of the structural factors estimated for the factor model. To the extent that is true, robustness of model performance to the new risk factor is clearly imparted. The raw stock universe to which the factor analysis is applied (the trade candidates) has some impact on results, as it does for the pair strategy. Nonetheless, with the matched stock universe, the factor model displayed desirable performance robustness compared with the pair model.

And what of nonfactor models? First, identify how factor risk is managed in the strategy, then extend the approach to the debt rating factor. For pair-based strategies, the obvious remedy is to homogenize all admissible pair combinations with respect to debt rating. That is, admit only those pair combinations of stocks in which the two constituent stocks have sufficiently similar debt rating. Then highly rated stocks will be paired with highly rated stocks, low rated stocks with low rated stocks, thereby avoiding bets on stocks that exhibited discordant price moves in response to concern over the debt factor. Many other aspects of debt rating and related issues would sensibly be investigated, too, including whether to employ position weights decreasing with debt rating, restrict low rated stocks to short positions, or an absolute veto on companies with very poor debt ratings.

As the research is pursued, an important question to answer is: What impact would have been seen on past performance of a strategy from incorporation of new modeling restrictions? It is all very well to congratulate oneself for finding the proximate cause of performance problems, to determine and implement prophylactic changes in one's modeling, but one also needs to know what impact on future performance (other than safeguarding when the factor is active in a negative sense) is likely to ensue. More extensive discussion of this subject in a broader context of performance disruption is presented in Chapter 9.

8.4 REDEMPTION TENSION

The pattern of redemption of a broad-based, long-only fund is perfectly "designed" to adversely impact a fast-turn, long–short reversion play. Selling off a long-only fund generates asymmetric pressure on stock prices—it is all one way, down. If the selling is broad based, and to some extent persistent, then the impact on spread positions can be only negative.

It is assumed that "broad based" means that a substantial portion of the stocks traded in the reversion strategy is affected—assume half. The long portfolio investment strategy, and hence current positions, is assumed to be unrelated to the reversion strategy: Approximate this by the assumption that the selling affects equally long and short positions of the reversion strategy. So, "all" affected longs reduce in value under the selling pressure of the redemption activity, and "all" affected shorts reduce in liability. On average there should be no net impact on the reversion strategy.

True ... to begin with. But what has happened to the universe of spreads in the reversion strategy? Those spreads in which both the long and the short are affected by the downward price pressure are essentially unchanged: Assume roughly proportional reductions in price. (In practice, stocks that have been relatively strong will be the initial focus of selling as the fire sale lieutenant seeks to maximize revenue—a one-sided, negative impact on a spread bet. The resulting price change will be larger for weak stocks when they are sold off, making the net result for a reversion-based book negative rather than nil. For this discussion, we continue with the optimistic assumption of zero net impact.) But for those spreads in which only the long or the short is in a stock facing redemption selling, the spread will change. Some will narrow, making money; some will widen, losing money. Still net nothing. But those narrowing spreads lead to bet exits—profit taking. The widening spreads continue to widen and lose more. Furthermore, continuing price reductions cause the spread model to take on new positions, which then proceed to lose money as the spreads continue to widen under continued selling. If the selling continues long enough—and for a large holding this is almost guaranteed—the natural trade cycle of the spread strategy will complete itself and those new trades will exit, locking in losses.

The picture can get even worse. When the selling is over, some stocks recover with a similar trend—as if there is persistent buying pressure. Who knows why that should be—reversion of relative value! For those stocks, new spread positions are entered: Remember, some previously entered positions finished on their natural dynamic and so the model is sitting and waiting for new entry conditions. Bingo, furnished as the stock price begins to reclaim lost ground. And the losing spread bet (now in the opposite direction) is made again.

High-frequency reversion strategies make lots of bets on small relative movements. Long-only fund redemptions cause price movements of a cumulatively much larger magnitude; the mechanics described in this section create the conditions for a blood bath for statistical arbitrage.

8.4.1 Supercharged Destruction

A large equity statistical arbitrage portfolio is perfectly designed to create, on liquidation, guaranteed losing conditions for other statistical arbitrage portfolios. Size matters because the sell-off of longs and the buy-in of shorts has to persist over the natural cycle of other players. If it does not, then initial losses will be reversed before existing trades are unwound; damage is limited largely to (possibly stomach churning) P&L volatility. Destruction, when it occurs, is supercharged because both sides of spread bets are simultaneously adversely affected.

In November 1994 Kidder Peabody, on being acquired, reportedly eliminated a pair trading portfolio of over $1 billion. Long Term Capital Management (LTCM), in addition to its advertized, highly leveraged, interest instrument bets, reportedly had a large pair trading portfolio that was liquidated (August 1998) as massive losses elsewhere threatened (and eventually undermined) solvency.

8.5 THE STORY OF REGULATION FAIR DISCLOSURE (FD)

Regulation "Fair Disclosure" was proposed by the SEC on December 20, 1999 and had almost immediate practical impact. That immediacy, ten months before the rule was officially adopted, is stark testament to some of the egregious behavior of Wall Street analysts now infamous for such practices as promoting a stock to clients while privately disparaging it, or changing a negative opinion to a positive opinion to win underwriting business and then restoring the negative opinion!

Activities eventually outlawed by Regulation FD had dramatic negative impact on statistical arbitrage portfolios in 1999. Most readily identifiable was the selective disclosure of earnings in the days before official announcements. Typically, favored analysts would receive a tip from a CEO or CFO. Analysts and favored clients could then act on information before it became public. If the news was good, a stock's price would rise under buying pressure from the in-crowd, statistical models would signal the relative strength against matched stocks, and the stock would be shorted. Days later when the news was made public, general enthusiasm would bid up the stock price further, generating losses on those short positions. Long positions, signaled when the news was disappointing, suffered similarly as prices declined, a lose–lose situation for short-term statistical strategies.

This pattern of behavior became so rife that many market participants yelled vituperation; the SEC heard and acted. Notable effects of the practice disappeared almost overnight.

Privileged information passed to analysts was not a new phenomenon in 1999. Widespread abuse of the privilege was new, or so it seems from the events just outlined. If abuse existed previously, it was not noticed. An interesting sidenote to the story is the effectiveness of analysts. Many analysts with star reputations for making timely and accurate forecasts of company performance became run-of-the-mill forecasters after Regulation FD was announced.

8.6 CORRELATION DURING LOSS EPISODES

An investor lament heard when enduring portfolio losses:

"Your results are [highly] correlated with those of other managers..." The implication here is that similar bets are being made, contradicting claims of differentiation through different methods of stock universe selection, trade identification (forecast model), and resulting trade characteristics such as holding period. Are the claims of differentiation false? Is the performance correlation coincidental?

Two distinct, broadly based portfolios of stocks traded with a reversion model are very likely to exhibit high coincidence of losing periods. When normal market behavior (patterns of price movement resulting from investor activities in normal times) is disrupted by an event, international credit crisis and war are two recent examples, there is a notable aggregate effect on stock prices: Recent trends of relative strength and weakness are promoted. During event occasions (and it is a near universal rule that "event" is synonymous with "bad" news), sell-off activity is always the result. Stocks perceived as weak are sold off first and to a greater extent than stocks that are perceived as strong (or, at least, not weak). This is the reverse of what is expected from a fund satisfying redemption notices—see Section 8.4. The implication for spread trades is blindingly obvious: losses. Regardless of the precise definition of a manager's strategy, tradable universe, or specific collection of active bets at the time, the common characteristic of all spread reversion bets, at a point in time, is that stocks judged (relatively) weak are held long and stocks judged (relatively) strong are held short. Mark to market losses are inevitable.

The magnitude of losses, duration of losing streak, and time to recovery will vary among managers, being strongly influenced by individual reversion models and manager risk decisions.

Any economic, political, or other happening that causes investors to become generally fearful instills a sell mentality. This has an unambiguous effect on all broad-based portfolios of spread positions on which mean reversion is being practiced. Unequivocally, performance turns negative; directional correlation of managers is high. Interestingly, numerical correlation may not be high. The magnitude of returns in negative performance periods can be quite different for distinct portfolios. There is nothing in the rationale of fear-based selling to suggest it should be ordered, evenly distributed across market sectors or company capitalizations, or in any other way tidy. Typically, a lot of untidiness should be expected. Hence, while managers should be expected to experience common periods of unusual losses, the correlation of actual returns in losing periods examined collectively and separately from returns in winning periods may be positive, negative, or zero.

Losing periods are always followed by winning periods, by definition. Extended intervals of market disruption punctuated by relief must, on the arguments given, create similar patterns of losing and winning intervals for spread reversion strategies.

And what can one say about relative performance of strategies during generally positive periods for spread trading? Less correspondence of returns for distinct strategies is to be expected. Returns are dependent on the precise dynamic moves exploited by individual models. There is no unifying force creating short-, medium-, and long-term dispersion followed by reversion that parallels the negative influence of fear. Exuberance is, perhaps, the closest to such a force, creating reversionary opportunities randomly and at large. But exuberance is less tangible than fear. It is less likely to induce common decisions or actions. Investment results for collections of managers will exhibit looser correspondence, greater heterogeneity than in periods unfavorable to reversion.

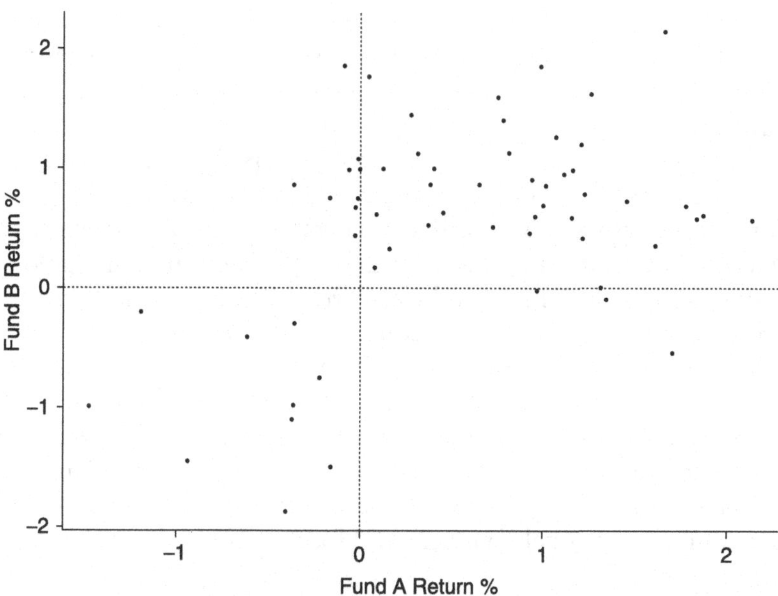

FIGURE 8.3 Monthly returns for fund A and fund B, illustrating the fallacy of correlation

Figure 8.3 illustrates the typical situation. Overall, fund A and fund B show positively correlated returns, with a modest correlation of 0.4. This correlation result is driven by the two quadrants where both funds are winners or both are losers (positive–positive and negative–negative) in which most trading outcomes lie. Within those two quadrants, the good and the bad strategy periods, correlation is negative: −0.19 in the bad times and −0.22 in the good times. This seeming contradiction, positive overall correlation but negative correlation in all dominant subperiods, is an example of the fallacy of correlation. Notice that the strength of the relationship between returns in the negative quadrant is actually lower, at 0.19, than in the positive quadrant (0.22), which is contrary to the general expectation described earlier. This example serves to illustrate a common theme described several times in the text, that while we can identify and characterize patterns—general or average—there is always variability to appreciate and contend with. Notice, too, that there are only ten data points in the negative quadrant, barely one quarter of that number in the positive quadrant. Thus, the correlation is less well estimated (there are fewer degrees of freedom, or pieces of information, in statistical parlance). And ten is a small number for estimating a variable relationship—isn't it?

It is not surprising that losing periods are experienced in common by different spread reversion strategies. The visibility of such correlation following two extraordinary years of market disruption is understandable. Understanding why the result is obtained is important: Attention is more likely to be rewarded if focus is shifted from contemplating the unexceptional coincidence of negative returns to examining where losses have best been contained. Attention should also be focused on the prospects for a resurgence of the drivers of spread reversion—on when and whether those drivers will reemerge strongly enough to create systematically profitable opportunities for managers (see Chapter 11). Here, there is real possibility of distinguishing future likely winners and losers.

CHAPTER 9

Trinity Troubles

Extinction occurs because selection promotes what is immediately useful even if the change may be fatal in the long run.

—T. Dobzhansky. 1958. "Evolution at Work." *Science* 1,091–1,098

9.1 INTRODUCTION

Beginning in early 2000, after nearly two decades of outstanding profitability, the returns of many statistical arbitrage managers collapsed to zero or worse. Some managers continued to generate excellent returns for two more years but they, too, ceased to perform starting in early 2002. The split into failures and successes in 2000 is an interesting historical point in the story of statistical arbitrage, demarcating a state change in high frequency reversion dynamics. Of greater significance because of its universal effect, the performance watershed in 2002 was considered by many to mark the death of statistical arbitrage as an absolute return generator, though there remained a few observers who discerned temporary structural problems and posited conditions under which statistical arbitrage would rise again. Coherent analysis was not abundant, investor patience was rarer even than that, and the latter became the real death knell as it led to investment withdrawal, leaving managers unable to meet payroll.

At the end of 2005 that was the dire state of statistical arbitrage; as an investment discipline it had become an unsaleable product. The year 2006 saw a resurgence in performance, vindicating those who had maintained that the performance collapse was explained by a multiplicity of factors, many of which were transitory. With the passing of those temporary disruptions to market coherence and consistently predictable security price dynamics, the likely possibility of a new productive period for statistical arbitrage was anticipated. We are now nearly two years into just that renewal.

In this chapter, we begin by examining several one-liners loudly touted as the cause of statistical arbitrage return decline. While each may have had some negative impact, it is unlikely that the combined effect has been more than 30 percent of historical return. A significant reduction, but not a coffin maker for the strategy. Next we expand the perspective to consider major developments in the U.S. economy and financial markets, describing the degree to which the impact on statistical arbitrage is transient. One perspective on 2003 is offered to set the context of the discussion.

Chapter 10 continues the theme of searching for reasons for performance decline and sources of a revival. The focus is on technical developments by large brokerage houses. Long-term impact on statistical arbitrage is certain, negative for many extant strategies but creating new opportunities with increasing use of the tools by large market participants (Chapter 11).

9.2 DECIMALIZATION

"The bid–ask spread reduced from a quarter to two cents has eliminated the statistical arbitrage edge."

Strategies may go by the same generic name, *statistical arbitrage*, but there are important distinctions that are critical to understanding performance disparities over the last few years as well as prospects for future performance. Starting in mid-2000, practitioners of high-frequency statistical arbitrage generally achieved a poor return, in many cases actually negative for much of 2000 and 2001. The declining bid–ask spread, from prices quoted in eighths to sixteenths to pennies, had an enormous negative impact on those returns. Furthermore, consolidation of floor specialists into the now five majors plus two small independents resulted in much of the intraday and day-to-day price reversion being internalized by those specialists. With research budgets, computers, and manpower resources exceeding most statistical arbitrage fund managers, and the unfair advantage of order flow visibility, it should not be a surprise that this has happened (see Chapter 10).

The statistical arbitrage edge was not eliminated. But the high-frequency opportunity was speedily removed from the public domain and monopolized by a favored few. Strategies with a holding period extending over weeks or months rather than days were largely unaffected by the change to decimalization. This dynamic characteristic of models was significant in explaining the performance split over 2000–2002 described in the opening section of the chapter. To those exploiting a reversion process fundamentally different from the high-frequency strategies, the contribution to profit of longer term bets from bid–ask spreads was fractional. There may have been some performance deterioration from poorer execution, but not outright elimination. To see this, let's take a look at some examples.

Consider an average stock priced at $40 per share. Suppose that the goal of a strategy is to earn 12 percent per annum, or 1 percent a month on average. With a holding period of two months, a reversion bet is expected, on average, to yield 2 percent, or 80 cents. The loss of the long-time standard bid–ask spread of a quarter, now reduced to a couple of cents following decimalization, can at most have eliminated one-third of the expected gain on a bet. Annual return is, therefore, reduced from 12 percent to 8 percent. This is a worst-case analysis that ignores the ameliorating possibilities available from trade timing tactics when trades can be made over several days—an option not available to higher frequency strategies. The actual impact of decimalization on the longer term strategy is more marginal.

The story doesn't end with decimalization, of course, as subsequent sections of this chapter show. Surviving the switch to decimalization, showing longer term statistical arbitrage strategies to splendid advantage in 2000 and 2001, did not help counter other structural changes that cumulated over 2002 and 2003.

9.2.1 European Experience

The European markets have been decimalized for decades yet high-frequency statistical arbitrage has developed successfully over the same time. The conclusion must be that decimalization itself is not a barrier to profit opportunity for the strategy; performance problems are created by changes in market structure which causes changes in temporal dynamics, disrupting the patterns that statistical arbitrageurs' models are built to predict. European markets are fully electronic, closer to the NASDAQ than the NYSE in that respect. Yet statistical arbitrage in all those markets failed to generate a return in 2003–2004. While it is possible that different causal factors explain the lack of performance in each of the "three markets," it is more likely that a common factor was active. What candidates might be implicated? The early part of 2003 was dominated by the Iraq war. But what of the final six months? And 2004? Economic news gradually shifted from the uniformly gloomy and pessimistic to generally optimistic, though the negative baggage of U.S. budget and trade deficits caused much consternation (on the part of commentators and professional economists). It is this change, and the commensurate changes in investor behavior, that caused the nonperformance of statistical arbitrage across markets and managers. In each of the markets, other market specific factors undoubtedly were also present.

9.2.2 Advocating the Devil

Having just argued that decimalization was not a significant factor in the reduction of return in statistical arbitrage, except for some high-frequency strategies, let's consider how the change could have been detrimental.

There is a lot of anecdotal evidence that liquidity within the day has changed directly as a result of decimalization: How do these changes relate to the day-to-day price behavior and trading volume for statistical arbitrage? *Prima facie* evidence, the decline of strategy return, is that market price patterns changed. How? In what ways? Can a logical connection be made between the observed changes, decimalization, and statistical arbitrage performance? Without considering the purported changed mechanics of intraday pricing, let us suppose the claim of change to be correct. What are the implications for daily price behavior?

Can a process be elucidated under which continuous trading from 9:30 A.M. through 4 P.M. will, *ceteris paribus*, generate daily price patterns structurally, notably, describably different depending on the size of the individual market price increment? If not, then systematic trading models evaluated on daily closing prices will also not exhibit distinguishable outcomes except to the extent that the bid–ask spread (at the close) is somehow captured by a strategy. In fact, simulations of many such models exhibited poor returns over 2003–2004. Either the observable, acknowledged, structural changes to price moves within the day resulting from the change to decimal quotes and penny increments led to change in the structure of end-of-day prices across time, or some factor or factors other than the change to decimalization explain the simulation outcome.

If a contrary observation had been made, then a plausible argument from decimalization to systematic trading strategy return could be constructed: If day-to-day trading shows positive return but intraday trading shows no return then price moves in reaction to trades eliminate the opportunity. The evidence to date neither supports nor contradicts such a hypothesis. It is much more likely than not that decimalization was a bit player in the explanation of statistical arbitrage performance decline.

Suppose that statistical arbitrage's historical performance was derived solely from systematic obtaining of the *consumer surplus* when spreads jump over a trade threshold and fills are obtained at or better than even that "excess." If decimalization reduces, almost to nil, the jump and, we might reasonably suppose (supported by experience), price improvement, too, then the expected return of bets similarly reduces to almost nil. This scenario is seductive until one realizes that it is little more than an elaboration of the bid–ask spread argument. The consumer surplus of the jump plus price improvement is quite simply the bid–ask spread (jump) plus price improvement. Unless price improvement was a major component of strategy return, this argument is reduced to dust.

9.3 STAT. ARB. ARBED AWAY

"Stat. arb. has not generated a return in two years. It's edge has been 'arbed' away."

This was heard with growing clamor as 2004 rolled on. But what kind of evidence is there for the dismissal? With nothing further than the observation of recent history offered to support the claim, we must suppose that the performance evidence was deemed sufficient proof. As such, the argument is refuted simply by looking a little further back. Over the 18 months from mid-1998 through the end of 1999, the strategy yielded almost nil return (factor-based models fared better than others), yet the subsequent two years yielded record rates of return.

An extended period of flat performance does not prove that patterns of stock price behavior, the source of return from systematic exploitation, have ceased to exist. Equally, there is no reason to suppose, solely by examining the numbers, that unusually large returns, or any particular pattern of returns, will be obtained as soon as a dry spell ends or, indeed, that there will be an end thereto. To understand what is possible, one needs to understand the nature of the stock price movements, the inefficiency that is exploited by the strategy, and how exploitation is achieved. To go further and posit what is likely requires one to make statements about various states of the world and to make forecasts (Chapter 11). Now, of course, the claim can also be refuted by looking forward from when it was made to today. Literally, there have been many happy statistical arbitrage returns.

9.4 COMPETITION

"Competition has eliminated the stat. arb. game."

It is tempting to dismiss this claim in an equally disdainful manner as in which it is presented. Leaving aside the implicit belittling of statistical arbitrage as a "game," those who practice it as such are playing roulette. Winning streaks are possible but when things go wrong the gamesters have no substance to fall back on. Rash action, desperation, and inglorious exit from the industry follow. For those who understand the drivers of their strategy and the subtleties of its implementation, shocks are survived through discipline and control.

Did competition eliminate risk arbitrage as an investment strategy? Quite! The dearth of opportunity during 2002–2005 was not because of a greater number of practitioners or increasing assets managed in the strategy, both of which preceded the return decline, but because of the structural change in the economy. As 2005 drew to a close, anticipation was already building that merger activity would increase, resuscitating the merger arbitrage business, with just a few months of consistently positive economic news. Increased participation in the business will have an impact on return as activity increases. The gains will be smaller on average; better, more experienced managers will do well if they discover and exploit the new patterns described in Chapter 11; neophytes, relying on traditional ideas, will have a more difficult time. Luck will play no small role.

What is the difference between merger and statistical arbitrage such that massive structural change in the economy—caused by reactions to terrorist attacks, wars, and a series of corporate misdeeds—was accepted as temporarily interrupting the business of one but terminating it (a judgment now known to be wrong) for the other? Immensely important is an understanding of the source of the return generated by the business and the conditions under which that source pertains. The magic words "deal flow" echo in investor heads the moment merger arbitrage is mentioned. A visceral understanding provides a comfortable intellectual hook: When the economy improves (undefined—another internalized "understanding") there will be a resurgence in management interest in risk taking. Mergers and acquisitions will happen. The game will resume after an interlude. There is no convenient label for the driver of opportunities in statistical arbitrage (although some grasp at "volatility" in the hope of an easy anchor—and partially it may be). There is no visceral understanding of or belief in how statistical arbitrage works; nothing to relate to observable macroeconomic developments; no simple indicator to watch. Beyond the visceral one has to think deeply. That is difficult and, hence, there is much uncertainty, confusion, and an unavoidable scramble to the conservative "Statistical arbitrage is dead." How is the Resurrection viewed, I wonder?

The competition argument deserves serious attention. Though there are no publicly available figures recording the amount of capital devoted by hedge funds and proprietary trading desks of investment banks to systematic equity trading strategies, it can be deduced from the remarks of clearing brokers; investors; listings in *Barron's*, Altvest, and so forth that both the number of funds and the amount of money devoted to the discipline increased greatly before 2000. An immediate counter to this observation as evidence supporting the competition hypothesis is that the increase in assets and number of managers has been taking place for two decades yet only with performance drought is a link to asset-class performance being made. Statistical arbitrage performance did not decline in tandem with a view of how assets/managers increased. The hypothesis was offered only to explain the *cessation* of performance for two years; an abrupt halt to a preceding excellent history. The hypothesis requires, to overcome the "convenient but not adequately matched to evidence" tag, an explanation of why the competition effect only recently, and dramatically, became apparent.

With the market in steep decline from 2000–2002, investors previously shy of hedge funds, including statistical arbitrage, increased allocations to alternative investment disciplines. Therefore, it may be argued that there was a step change (increase) in investment in statistical arbitrage in 2002.

But ... it did not all happen at the beginning of 2002, did it?

What other evidence, besides assets invested in and number of practicing managers, can be sought to support or discredit the competition hypothesis? The superficial argument considered thus far, increased attention correlated with poor performance, is barely an argument at all; it is really only a coincidence of observations with correlation taken for causality and no explanation of how a causal mechanism might be expected to work. The simplest scenario is "many managers were competing to trade the same stocks at the same prices." Even an approximation to this would be revealed to participants through liquidity problems—unfilled trades, increased slippage on filled trades, for example. Evidence to support such an explanation has not been widely claimed.

If competition during trading eliminates the "consumer surplus" and price improvement (historically part of statistical arbitrage return) then once again the effect should be visible in end-of-day closing prices. The fact that many bets continue to be identified with a substantial consumer surplus component belies the argument. The reduction in number of opportunities is directly related to volatility, which may very well be reduced in some part by greater competition among a larger number of statistical arbitrage managers. That still leaves the important question: Why is the sum total of return on the identified opportunities reduced to zero?

Let us accept that competition in systematic trading of equities has increased. There is no evidence, notwithstanding performance problems, to support concomitant increase of market impact, and consequently no evidence that greater competition is the major cause of the decline of statistical arbitrage performance.

9.5 INSTITUTIONAL INVESTORS

"Pension funds and mutual funds have become more efficient in their trading."

Three years of market decline pushed many institutional investors to enormous effort to reduce costs and, hence, losses as return generation eluded most; transaction costs were prime targets. The argument is that reversion opportunities historically set up by the large block trades disappeared as traders of those blocks became smarter with their trading. "Fidelity has been a 'VWAP* shop' for several years" is frequently heard as shorthand for the argument. Once again, we must note that these changes did not happen overnight. To the extent that such changes have been a factor in statistical arbitrage performance decline, it is confounded with other changes. Assessing the impact appears to be an insurmountable practical problem. Certainly, institutional investors are major users of the trading tools described in Chapter 10 and a substantial impact—negative for statistical arbitrage—is undoubted.

9.6 VOLATILITY IS THE KEY

"Market volatility ticked up—isn't that supposed to be good for stat. arb.?"

From the beginning of 2002, people began searching for explanations for the lack of return from statistical arbitrage strategies. Many managers had experienced a meager year in 2001 though many others had a good year. But shortly into 2002 managers everywhere (almost) were experiencing poor performance. The decline in market volatility was dragooned into service as "the" explanation for the lack of statistical arbitrage performance. Impressively quickly the volatility explanation became Antilochus' hedgehog, a single "big" idea. Combined with investor pleading for silver bullet solutions to the performance drought, observers might be forgiven for wondering if they had entered a film set! This chapter and the sequel stand as explanation in the mode of Antilochus' fox: many "small" ideas.

*Volume Weighted Average Price.

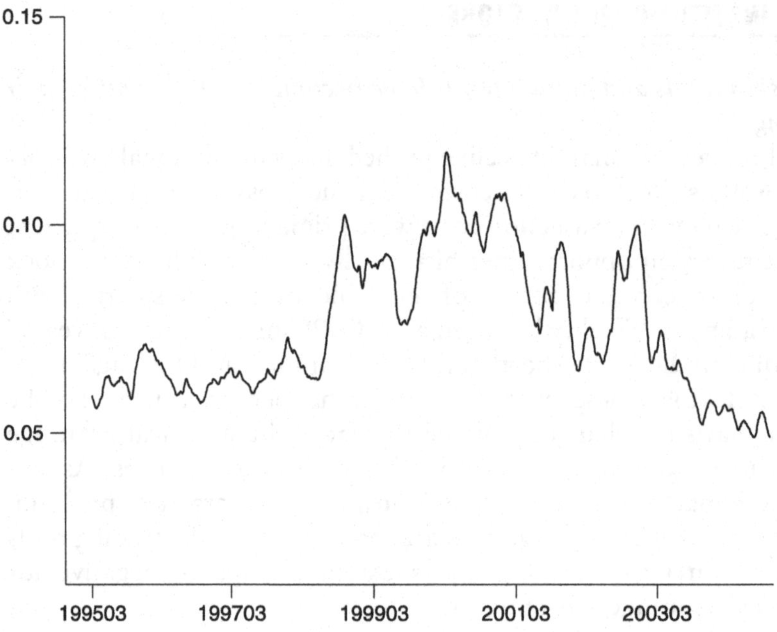

FIGURE 9.1 S&P 500 within industry average local pairwise volatility

Spread bets exploit relative price movement between stocks. It is *interstock* volatility that is critical for performance; and interstock volatility, while constrained by market volatility, is not a simple function of it (see Chapter 6). Often interstock volatility moves in the contrary direction to market volatility. In the third quarter of 2003, interstock volatility declined to a record low, even as market volatility increased. The decline continued through 2004.

Just as interstock volatility is not simply related to market volatility, so the level of interstock volatility is also not simply related to strategy profitability. Only a small proportion of the total volatility is systematically exploited by spread models; large changes in the level of interstock volatility are observed to have only a small impact on the magnitude of strategy return (except when volatility decreases to extremely low levels as seen in 2004 and except for more sophisticated models which capture more of the raw volatility) a larger impact is observed on the variability of the return. Both relationships are well demonstrated by the contrasting market conditions and strategy performance in the first and third quarters of 2003. The record low level of interstock volatility in quarter three preempted consistent profitability for the first time that year; volatility was 20 percent higher in the first quarter, yet performance was wholly negative.

In September 2003, interstock volatility declined to a record low level, yet reversion opportunities were rich. Statistical arbitrage strategies generated a 1 percent return *unleveraged* in two weeks.

Unfortunately the trading activity of Janus, funding $4.4 billion of redemptions precipitated by the firm's disclosure of participation in mutual fund timing schemes in contravention to statements in fund declarations, disrupted price relationships in the second half of the month. A corollary of the Janus story (the redemption detail was published in the *Financial Times* on Friday, October 10) is that almost certainly more "disruption" should have been anticipated for October. Morningstar (*Financial Times*, Thursday, October 9) advised investors to reduce or eliminate holdings in mutual funds from Alliance Capital and Bank of America, as managers of those funds had also engaged in timing schemes.

9.6.1 Interest Rates and Volatility

With very low interest rates, the value of a dollar a year from now (or five years or ten years) is essentially the same as the value of a dollar today. Notions of valuation of growth are dramatically different than when higher interest rates prevail, when time has a dollar value. With such equalization of valuations and with discriminatory factors rendered impotent, volatility between similar stocks will decrease. Higher interest rates are one factor that will increase stock price discrimination and increase the prevalence and richness of reversion opportunities. The process of increasing interest rates began at the end of 2004. The Federal Reserve raised rates in a long, unbroken sequence of small steps to 5 percent, and statistical arbitage generated decent returns again starting in 2006.

Volatility has not increased since 2004. Indeed, it declined further to record low levels. *Can* volatility rise to historical levels? Absolutely. But the developments cited here, the rise of VWAP, participation and related trading, and the trading tools described in Chapter 10 strongly suggest that it would be foolish to bet on it.

9.7 TEMPORAL CONSIDERATIONS

The foregoing analysis is entirely static in focus: decimalization, competition, and so forth are not apparent for any individual trade (on average) as evidenced through slippage calculations. (If you prefer, conservatively, there is impact but the magnitude is insufficient to cause the outcome of zero return.) Statistical arbitrage is not simply a collection of individual trades either in distinct stocks, or pairs of stocks, or more general collections of stocks. It is a collection of linked trades related over time. The temporal aspect of the trades is the source of strategy profit; trades at a point in time are the means by which the opportunity is exploited. The foregoing argument demonstrates that competition has not inhibited the ability of managers to exploit identified opportunities. But has competition, or decimalization, or something else altered the temporal structure, the evolution, of prices such that identified patterns ceased to yield a positive return? Did short-term stock price structure change such that systematic trading models were reduced to noise models? If so, can the progenitor forces driving the evolution be identified? Was decimalization or competition influential? Were they active agents, catalysts, or simply coincidental elements? Are they still active factors? If there are other factors, how have they caused structural change? Is the process over? Is there a new stable state—now or yet to be—established, or will the status quo be restored?

Once again, the starting point is the observation that statistical arbitrage strategies, historically yielding good returns, did not generate a decent positive return in most cases in at least three years through 2005. Many such strategies lost money in one or more of those years. The foregoing analysis has considered widely posited hypotheses that performance was crowded out by changes effective at the point of trade placement and shown them not to reliably or reasonably provide an explanation of the observed pattern of trading and opportunity set identification. What about the hypothesis that from the same changes—decimalization, competition, or other unidentified factors in conjunction therewith—has arisen a change in the temporal structure of stock price behavior such that previous models that once had identifiable and systematically exploitable forecast power now have none? What would one expect to see from systematic trading strategies if the signal component of the model was reduced to noise? Bets would be placed that, on average (and hence in the aggregate), have zero expected return. With random elements and varying degrees of expertise on the part of managers executing trades, systematic strategies would yield zero to negative performance (transaction costs providing a negative bias to the zero expected return of raw trades).

On the face of it, three years of essentially flat return by the "class" fits the hypothesis. Is there specific evidence that one might look for to determine if the hypothesis is merely a fit to the observation of overall performance or drove the result? If a model has no forecasting power, then forecast returns should be uncorrelated with actual returns for a collection of identified bets. The evidence of one fund is known in some detail. Unequivocally, the evidence refutes the hypothesis. For the majority of trades over the three years, the correlation between forecast return and achieved return is positive (and statistically significant). Many of the trades generated a positive return, though, on average, a lower return than in previous years. The minority of losing trades made bigger losses. This is the crux of the performance problem for many statistical arbitrage managers, though there are some additional elements that contribute to the story and have implications for prospective results. In somewhat handwaving terms, one can characterize the situation as follows: The signal remained present (generally high percentage of winning bets); it was somewhat weaker (lower rate of return on round-trip, completed bets); the dynamic became erratic (variably longer or shorter holding periods); and the environmental noise increased (higher variance on losing bets, proportion of winning bets, location of consistent reversion).

An archetypal example of a wave function, a ripple in a pond, a sinusoid for those of a technical bent, helps illustrate these components, their contribution to bet performance, and the implications of the changes discussed (Figure 9.2).

To begin with, remember that noise is good: Suppose that observations were simply scattered about the pure signal with an additive random element, that is $y_t = \mu_t + \epsilon_t$ with $\epsilon_t \sim [0, \sigma]$. Then a large noise variance σ would generate series such as Figure 9.3 compared to low noise Figure 9.4.

The same *signal* in Figure 9.3 admits a greater return through understanding of the signal (model) and the impact of the noise

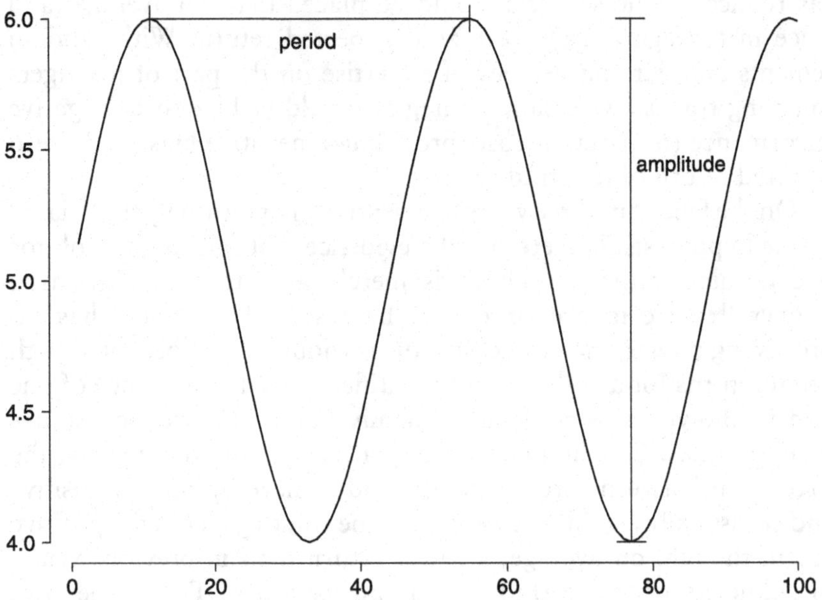

FIGURE 9.2 Sine wave

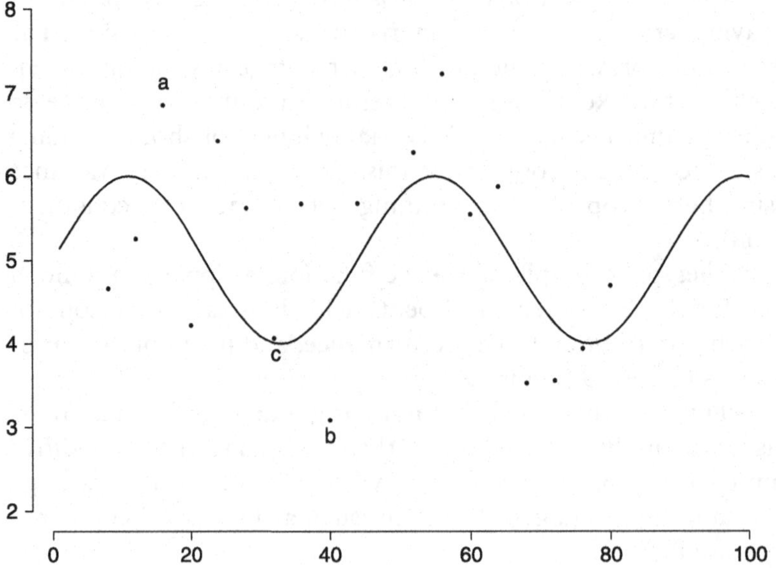

FIGURE 9.3 Spreads with underlying sine wave signal

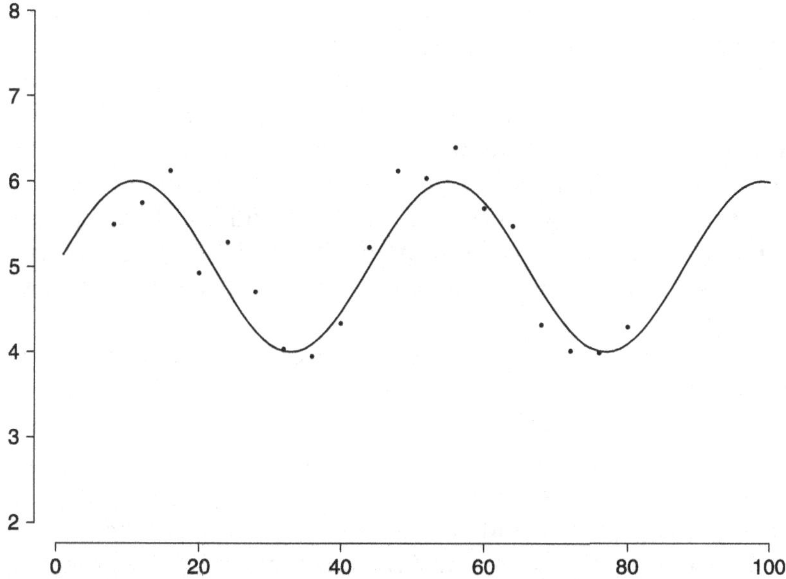

FIGURE 9.4 Low volatility spreads with underlying sine wave signal

component. We know where the signal is going and we know how much variation about the signal may be preempted by the noise. This leads one immediately to modify a simple exploitation of the *signal* forecast, to one that also exploits knowledge of the variation anticipated from noise. Rather than exiting a bet when the model forecasts 0, one identifies an interval around the 0 and waits for an outcome much to one's favor that will occur according to the distribution of the noise. This phenomenon is known as stochastic resonance (see Chapter 3). Enter at *a*, exit at *b* and not at *c*. There will be gains and losses (missed trades, opportunity costs) compared to the "exploit the signal" model; on average, with good calibration, one will gain. Clearly, the trade-off of a noise gain against opportunity cost (capital not available for another new bet) is different—there is much less potential noise gain—in Figure 9.4.

In 2003–2004, much commentary was made about the inaction of institutional money managers; "sitting on the sidelines" is an apt description of both the lack of commitment to active decision making and the wait-and-see posture adopted in response to a backdrop of poor economic and political news, not to mention an unprecedented three-year rout of equity markets. This reduced level of investing activity makes an impact on reversion structure (in prices of similar stocks largely owned by institutions and followed by professional analysts) in several ways. Foremost is the slowing down of the pace of reversion—an increase in the period of oscillation in the wave function in Figure 9.2. Pull apart the ends of the wave in that figure . . . and . . . things . . . move . . . more . . . slowly. If nothing else changed, this dynamic shift alone would dramatically reduce strategy return: If it takes twice as long for a move to occur, then return from that move is halved.

The practical impact is larger than the archetype suggests; return is reduced by more than half, because there are compounding factors. Probably the most significant factor is the ability of a modeler to recognize the occurrence of the change of dynamic and modify models appropriately. Working with mathematical archetypes, one can immediately identify and diagnose the nature of a change (of period in a sinusoid, for example). With noisy data in which the signal is heavily embedded in extraneous variation, the task is enormously more difficult. In many cases, the change is swamped by the noise, which itself may exhibit altered distribution characteristics as we are seldom so fortunate that change occurs one factor at a time or in some other conveniently ordered manner, with the result that detection of the change takes time—evidence has to be accumulated. Delays from identifying a change in dynamic result in reduced return from systematic signal exploitation.

The process of change adds yet another level of complexity and another source of drag on return. Implicit in the preceding discussion has been the notion of instantaneous change from an established equilibrium to a new equilibrium. Rarely is that how systems develop. Much more prevalent is a process of change, evolution. Such a process may be more or less smooth but with a signal embedded in considerable noise, such a distinction is practically moot. Whether one posits a smoothly changing signal or a series of small but discrete, differentially sized, changes as the path between equilibria the outcome is largely the same: further reduction in return as one's models are shaped to reflect a new signal dynamic. Overshoot and elastic-like rebound add yet more volatility and, hence, uncertainty.

Responding to changes in the structure of prices (and other metrics where they are used) is one of the more difficult tasks facing a modeler. Unlike physical processes such as the motion of a pendulum subject to friction and wear, or chemical processes where impurities affect conversion properties, and so forth, there is no underlying theory to guide one's model building for distortions in the relative pricing of equities and the impact on reversion mechanics. It is not a mechanical process at all. It only appears "mechanical" in "normal" times when disturbances are infrequent and have limited duration impact such that a statistically regular process is observed, exploited, and return streams are pleasing to investors. Better models are inherently adaptive so that a change in, for example, the volatility of a stock price, if persistent, will automatically be identified and the model appropriately recalibrated. A key requirement for effective operation of adaptive models is the persistence of a new state. When uncertainty is reflected through a succession of changes (in, say, volatility) first in one direction then in another, an adaptive model can fail ignominiously as it flails hopelessly against waves of varying magnitude and direction. In such circumstances, rigidity is a better bet. A modeler, observing markets, the induced adaptations in his models, and the practical results of trading, ought to develop a sense of when rigidity is a better vessel than continual adaptation. A difficulty here is in choosing a model calibration at which "to be rigid" and when to apply or remove rigidity restrictions. Left to the informed judgment of the modeler alone, this is an art. Some modelers are extraordinarily talented in this respect. Most are hopeless. The temptation to tinker when things are going poorly is, for many, irresistible, especially for those who do not have a firm understanding of the process being exploited by the model or how the model's performance is affected by violations of assumptions.

With a keen appreciation of the high probability of failure from tinkering, and a realization of the nature of the difficulties besetting a model (whether one is able to construct a coherent explanation of why investors are behaving in ways that cause the aberrant price patterns), a good modeler looks to build automatic monitoring systems and to design robust feed forward and feedback mechanisms to improve models.

Starting with an understanding of the signal that a model is exploiting, a monitor scheme is constructed to repeatedly ask the questions, "Is the data violating model assumptions?" "Which assumptions are not being met?" One criterion, episodically informative and efficacious, is the set of conditions in which a model is known to perform poorly: When a model is observed to be adapting frequently, back and forth or repeatedly in one direction or cumulatively by too much, a modeler's intervention is desirable.

Feedback mechanisms form the bread and butter of an adaptive model. If stock price volatility is observed to have increased by a sufficient margin, recalibrate the model. Repeated feedback adjustments outside of the range (frequency, magnitude, cumulative impact) in which the model is known to perform adequately are a warning sign for the monitoring process. Feed forward mechanisms are not typically available to managers: We cannot change environmental conditions to coerce a desired change in patterns of stock price development. Larger funds are reported to engage in activities of this sort, using, among other schemes, fake or phantom trades—real trades to reveal to other market participants the manager's supposedly desired moves, buy IBM for example; then wait for those others to follow the lead, buying IBM and pushing up the price; then the fund sells its holding—its original intention—at prices more favorable than before the faking operation. Managers do not admit to such activities. Technological developments may be providing more opportunity for secret tactical trading: See Chapters 10 and 11.

We entered this discussion of monitoring for and adaptation to structural change by considering the impact of relative inactivity by institutional money managers. Such inactivity, or lack of enthusiasm generally, in an environment exclusive of sources of fear, reduces the (reversion) opportunity set in a second way. Returning again to the archetype in Figure 9.2, a lack of energy in a signal translates into a smaller amplitude (peak to trough range). In terms of a spread between prices of similar stocks, the action of prices moving apart either because of nonspecific, local drift or in reaction to investors pursuing a particular thesis, movements are muted in magnitude as excitement is constrained by the prevailing environmental condition of wariness (and in some cases lethargy) borne of a belief that as things (market themes, prices, activity) are generally not changing rapidly, there is no sense of likely opportunity loss to be incurred from taking one's time.

As institutional money managers saw returns diminish (or disappear, or worse, for three years) many altered their trading tactics as part of an attempt to reduce costs. Where traditionally (large) trades were simply handed over to a block trading desk for execution, managers began working trades directly, reducing (it is claimed) brokerage charges and slippage costs quite dramatically. It is suggested that this change (a) contradicts the hypothesis that institutional money managers were less active over 2002–2004, and (b) contributed to the lack of statistical arbitrage performance.

We do not have figures from which to draw evidence to confirm or deny (1) and in any case the episode is over, so we will leave it. Regarding (2), the evidence is to the contrary. If managers responsible for substantial volumes of trading have changed trading tactics to become more efficient or to be more proactive in reducing market impact or to reduce slippage of their trading, what would one expect to see in statistical arbitrage focused on stocks largely held by institutions?

To the extent that block trading activity is a generator of interstock dispersion (creating openings for reversion bets), a shift away from block trades to more intensively managed, smaller trades with greater control of slippage would reduce the reversion opportunity set. We would see a diminution in the average richness of reversion signals: A manager moving capital into, say, drug stocks would cause, for example, the Pfizer-Glaxo spread to move by a smaller amount than under the less demanding block trade approach. With a smaller initial dislocation, the amount of reversion is reduced. It is also possible that the number of economically interesting reversion opportunities would be reduced, though with other sources of market price movement present it is not obvious that this would be a significant effect. By itself the reduction of average reversion per bet would reduce strategy return. However, other effects of the trade tactic change make that conclusion premature. With managers more directly active, it is likely that their own trading decisions that act to enforce reversion would increase the pace at which reversion occurs. Faster reversion works to increase return (assuming that there are enough reversion opportunities to fully employ capital). While it is possible to argue that managers will act in the manner described to reduce average dispersion, hence potential reversion, yet not act as robustly to correct a mispricing when seeing one, it is unreasonable—you cannot have it both ways.

Is there any evidence to support the outcome that greater micro trade management by institutional money managers implies? It is abundantly evident that the pace of reversion *slowed*, not accelerated, during 2002–2004. The evidence on richness of reversion opportunities is more equivocal. There have certainly been periods where interstock volatility has been at a record low level—March of 2003 stands out as a period when most stocks moved in close unison. But here the diminished volatility was the result of global security concerns; it had nothing at all to do with money managers watching the dollars and cents.

9.8 TRUTH IN FICTION

The accusations flung at statistical arbitrage as reasons for its poor showing each include a truth. Each of the causes posited have had a negative impact on the size of return that statistical arbitrage models are able to generate. But in sum these slivers of return amount to no more than 30 percent of the return "normally" (that is, before 2000) generated. We are impelled[1] to search for a wider rationale for the performance collapse in statistical arbitrage. Hints are apparent in the previous section on temporal dynamics. Now we can be more explicit.

9.9 A LITANY OF BAD BEHAVIOR

Table 9.1 lists a series of events spanning the two years 2002–2003, each of which had a significant impact on business practices and financial market activities greatly in excess of "normal" change. Most events were negative, in that shock, disgust, and not a little horror characterized the reactions of many.

The first few months of 2002 subjected people to an unprecedented (?) series of appalling revelations about the activities of business leaders and opinion-leading Wall Street personalities. These

[1] Stephen J. Gould, 2002, *The Structure of Evolutionary Theory*, provided the phrase "impelled to provide a wider rationale for" shamelessly borrowed for my purpose here.

TABLE 9.1 Calendar of events

Date	Event
December 2001	Enron
January 2002	Accounting scandals, CEO/CFO malfeasance
	Wall Street research: lies, damned lies, and millionaire analysts
August 2002	Corporate account sign-off
October 2002	Mutual fund retail investor panic
November 2002	SARS
March 2003	Iraq war
	Dividend tax law revision
	NYSE/Grasso compensation scandal
October 2003	Mutual fund market timing scandal
December 2003	Statistical arbitrage investor flight

events delivered emotional punch after punch to a populace still in a deep sense of shock following the terrorist attacks on the United States on September 11, 2001. Unsurprisingly, the financial market parallel to the shifts in macroeconomic activity was structural change in the relationship of stock prices on a huge scale and with the effects of one change merging into the next. No rest. No respite. Continuous turmoil.

As 2002 was drawing to a close, the SARS (severe acute respiratory syndrome) scare dealt another blow to international air travel with impact on international tourism. In late 2004, the World Health Organization attempted to raise consciousness about Asian bird flu, forecasting that the serious outbreak in Asia threatened to become a worldwide epidemic that could kill 50 million people.[2]

As of now there is no sense of panic, even urgency on the part of political leaders or populations. Little heed at all seems to have been taken. That reaction is astonishingly different to the reaction to SARS just two years earlier. Can it be that people have become bored with scare stories?

Along with SARS the world watched the inexorable buildup of U.S. military forces in the Gulf of Arabia. Would the United States

[2]January 13, 2007: Thankfully no epidemic has occurred, but concern remains as deaths of farmers in China and elsewhere in Asia continue to be reported.

invade Iraq? The watching and waiting continued through March 2003 when the United States did invade. During the three weeks of "active hostilities,"[3] the markets ceased to demonstrate evidence of rational behavior on the part of investors. If television showed pictures of explosions with reports of problems for American troops, no matter how local a battle was being described, markets moved down. If television showed pictures of American military hardware on the move or cruise missiles raining down on sand, markets moved up. So much for the sophistication of the most sophisticated investors in the world in the most sophisticated markets in the world. If this were not real, with real implications for livelihoods, it would be truly laughable.

In the last quarter of 2003, evidence of maturing investor response to yet more bad news was readily visible in the reactions to (then) New York attorney general Spitzer's revelations of illegalities on the part of mutual funds. No wholesale rout of the industry ensued. Investors calmly withdrew from the shamed funds and promptly handed over monies to competitors. The surprise value of further bad faith activities on Wall Street (a convenient, if amorphously uninformative label) was met with rational analysis and not shocked, unthinking panic. Doubtless a rising market for the first time in three years had a powerful aphrodisiac effect.

The sequence of shocking events, each starkly disgraceful individually, is an appalling litany. Added to globally disruptive events (war, health scare) there were two years of uninterrupted instability in the financial markets.

How does market disruption affect the process of relative stock price reversion? Figure 9.5 extends the previous view of an archetype spread (Figure 9.2) to cover a period of disruption. A spread is pushed unusually far out of its normal range of variation by undisciplined investor behavior but after the cause of the panic is over, or as panic reaction dissipates and discipline is reestablished, the spread resumes its normal pattern of variation.

[3]The occupation of Iraq has been an unending run of active hostilities. The hostility of insurgents remains virile; in early 2007 President Bush directed 20,000 additional U.S. troops to be sent to Baghdad. However, events in Iraq have long ceased to have noticeable impact on financial markets.

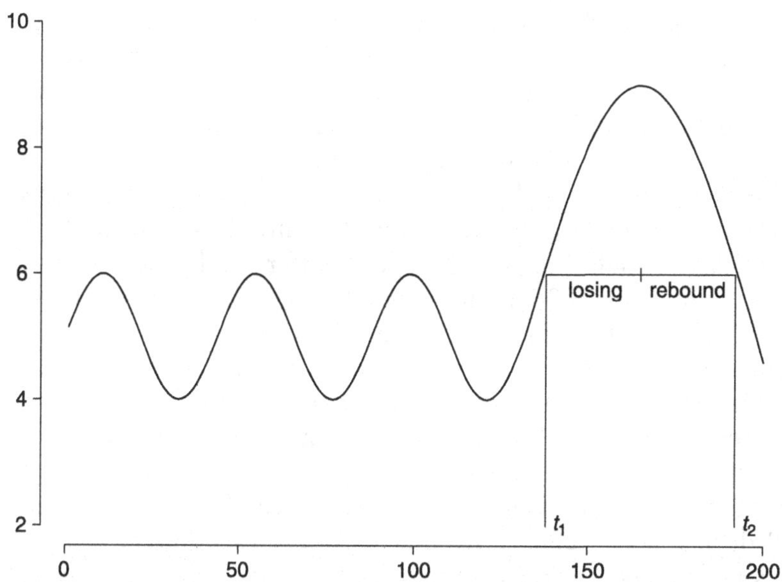

FIGURE 9.5 Sine wave with temporary perturbation

The model has at time t_2 a different assessment of the normal behavior of the spread than it does at time t_1 (just as it potentially has a different view at all times but consideration of just a few points is sufficient for a clear illustration of the temporal development, the evolution of views encapsulated in an adaptive model). Depending on the model's look back, the rate of information (data history) discounting, the projection of future pattern—focus on the mean for this analysis—will vary. A shorter look back (faster discount) will project a higher mean (also greater amplitude and longer phase) and will signal exit too early, which will lock in a loss on the trade. A longer look back (slower discount) will generate standard profit (in this example) but over an extended duration, hence, much lower return. Nowhere will gains be accelerated. Hence, return unambiguously must decline (in the absence of intervention).

In the next section we offer a particular view of the unfolding psychology of market participants over 2003. This is intended as a serious analysis (though necessarily of limited scope) of the nature and causes of observed changes in market participants and the resulting effects on security price development. It builds on the description of events just given, the goal being to reveal the extent of the changes in the U.S. macro economy, polity, and financial markets. Section 9.11 describes how such tectonic shifts in environment affect quantitative models and what modelers can do to manage the shifts. Section 9.8 foreshadows that discussion. There are no simple one-liners in these descriptions and analyses. The world is not so obliging. Statistical arbitrage did not generate a return in three years for multiple reasons, differentially across strategies, differentially across time. What happened in 2005 is not what happened two years previously. Attempting to explain the failure by citing one or two easily spoken reasons, as if trying to pin a tail on a donkey, is unhelpful. It simplifies the complexity beyond the point of understanding, leading to misunderstanding. It eliminates any ability to sensibly postulate what is realistically possible for statistical arbitrage. With the return of statistical arbitrage performance since 2006, the criticisms voiced during 2003–2005 have magically been forgotten. Chapter 11 tells how the statistical arbitrage story is about to write a new, positive chapter.

9.10 A PERSPECTIVE ON 2003

Trading intensity in the first quarter was low because of investor hesitancy in taking positions as the United States prepared to go, then went, to war. Following the three weeks of active hostilities, investor activity in the markets was at times tentative, skittish, manic. As early as last quarter 2003, it was possible to see, swamped in short-term variability, a trend of steady improvement in tenor: Uncertainty had decreased, conviction and the willingness to act thereon had increased; companies were again investing strategically, implementing long-term plans rather than substituting short-term holding actions; individuals were increasingly leaning toward optimism on prospects for employment opportunities and stability. Crucially, the pervasive sense of fear, not at all well defined or articulated but palpable from late 2002 through May 2003, was gone.

The summer recess both interrupted and contributed to the post-war recovery. Distance in time provides perspective; a change in routine, vacation, encourages reflection. The doldrums of summer (interstock) volatility were lower than had been seen before, partly because of the low pre-summer level and partly because of the clear need for people to take a break. Two critical changes in perception occurred.

The tragic, daily loss of servicepeople's lives in Iraq impinged on the general populace's consciousness with the interest, intensity, and indifference of the latest rush hour traffic accident: It is there, it is unfortunate, but it is reality. The U.S. economy was discussed in encouraging terms of growth and stable employment. Deflation had resumed its traditional role as a textbook scenario. The number of people who understand or care about government deficits and the implications—until they occur—is tiny. Such broad changes in perception have an indelible imprint on financial markets.

Transition from war edginess and economic gloom to war weariness (dismissal) and the excitement of economic potential, opportunity: Market price behavior in the latter part of 2003 reflected investor fervor, alternately hesitant, rushed, somewhat erratic; generally untidy, undisciplined.

9.11 REALITIES OF STRUCTURAL CHANGE

The complexity of the process of change is revealed in the mixed signals of market condition tracking and prediction models. From March 2003, early in statistical arbitrage's performance desert, these models simultaneously indicated both a shift from negative to positive bias and continued negative bias. Unique in more than a decade, the schizophrenic indicators revealed a market structure in which that structure is mixed up, unsettled, and in transition. The evolution of the indicators, if one were to imbue them with some life force, engenders the impression of a relentless striving for equilibrium, increasingly confident each month.

Adapting to the changes in market price behavior that reflect the enormous changes in perceptions, concerns, assessments, and ultimately, actions of market participants is extraordinarily difficult. For models designed to exploit an identified pattern of behavior in prices, the task can be impossible (if the exploited pattern vanishes) and often models simply "do not work" during market structural change. Evolution and adaptation are possible in better models but large, abrupt shifts and repeated shifts are immensely difficult to manage well.

Statistical arbitrage models have no special protection from the impact of market upheaval. The fact that performance diminished to little more than money market returns in conditions, known with 20–20 hindsight to be quite unfavorable for reversion exploitation (lack of consistent behavior untrammeled by decisions borne of panic), is a testament to the careful construction of traded portfolios, strict adherence to model application where signals are detected, and concerted focus on risk analysis. Critical to risk control is an understanding of the process exploited by the model: in short, why the model works. Reversion in relative prices of similar stocks did not evaporate—models systematically identified opportunities and trading has routinely exploited those opportunities. Reversion did not evaporate. The environment in which reversion occurred was changed, transforming how reversion is identified.

Changes of state are typically unrewarding periods, even negative, for statistical arbitrage. Models, good models, are crafted carefully to adapt to changes in important characteristics of market price behavior pertinent to model predictive performance. But no matter how hard model builders try, diligence cannot compete with good fortune when structural changes occur. It behooves us to admit that if we can avoid losses during structural change, while recrafting models to encapsulate the new structures, then we have done well.

9.12 RECAP

At this point we have concluded that a third of the historical performance of statistical arbitrage may have been eliminated by market developments during 2000–2002, changes that will not be reversed. The loss of the bulk of the historical return in 2002–2003 was the result of a series of massive disruptions to the U.S. economy, the ramifications for statistical arbitrage having (most likely) been felt in their entirety by some time, probably early, in 2004. The frequency of disruptions has been greatly reduced; though there continue to be market effects as the structural changes play out, the impact on statistical arbitrage is no longer significant. Stock-specific events continue to occur, two examples in 2004 being the withdrawal of the drug Vioxx by Merck and the investigation of Marsh McLennan by (then) New York attorney general Elliot Spitzer. Extraordinarily low volatility coupled with high correlations is a major limitation on what can be made from reversion plays. Correlations have now decreased, increasing the short-term reversion opportunities. Working to keep correlations higher than historical norms is the growing use of exchange traded funds. As investors shift to ETFs, "everyone becomes a *de facto* indexer." Volatility will continue to be constrained by the widespread use of sophisticated trading tools (Chapter 10). But that very same causal factor is contributing to the renaissance of statistical arbitrage by creating new kinds of systematic stock price patterns, as elucidated in Chapter 11.

CHAPTER 10

Arise Black Boxes

Felix qui potuit rerun cognoscere causas.
Happy is he who can know the cause of things.

—Virgil

10.1 INTRODUCTION

Having invented the pairs trading business two decades ago, Morgan Stanley was at the forefront of the creation of a new business in the early 2000s; a less risky, more sustainable business, which, in a wonderful example of commercial parricide, has systematically destroyed opportunities for old-line pairs trading. Algorithmic trading was born. Huge order flow from institutions and hedge funds, much of which is electronically matched in house, provided multiple opportunities for bounty beyond the expected brokerage fees. Combining the insight and knowledge learned from proprietary trading (beginning with the classic pairs trading business) with analysis of a warehouse of order flow data, Morgan Stanley and other brokers built trading tools that incorporate models for forecasting market impact as a function of order size and time of day, moderated by specific daily trading volume stock by stock.

Recognizing that there was an enormously lucrative opportunity hanging on simple to use, automatic trading technology that did not systematically incur slippage, brokers elected to offer the tools to clients. It was a masterfully timed decision. Coming as new statistical arbitrageurs were appearing with abandon, vendors were able to seduce those whom their tools would eventually help destroy, along with existing clients thirsting for any new edge that had the promise of lower transaction costs or marginal improvements in execution price. The genius of the business was compounded as the institutional and statistical arbitrageurs' order flow provided an ongoing feast of data for the data miners whose voracious appetite for such cannot be sated.

Patterns of transaction volume by stock, by day of the week, by time of day, and by current day's trading volume were constructed from the mined data. The mere ability to predict with measurable efficacy how much would be given up from current price to buy or sell a specific number of shares in a fixed period was a stunning development to traders. Hedge funds had for years made their own attempts; using their much less rich data than broker archives it is unlikely their achievement matched the brokers' success. Regardless, an edge was eliminated.

Fitting logistic-type models to order flow and fill data quickly produced the first generation of models, allowing traders to obtain quantitative answers to frequently faced, urgent questions:

- How much will I have to pay to buy x thousand shares of XYZ in the next half hour?
- How much will I have to pay if I wait the remainder of the trading day?
- How much can I sell of XYZ in one hour keeping the impact to k cents?

An unadvertised beauty of these tools is the self-propagating nature of the opportunity set. As traders switched to the technology, a new set of order flow information was presented to and collected by vendors. Now it was possible to examine the trading of both the impatient "pay up and get it done" and the relaxed "wait and see" players. Models of client profiles, built automatically from the client order flow, trading tool configuration, and fill/cancel–correct records practically generate themselves. With the ability to gauge how much a client would be willing to pay for a fill, and estimates of how long it would take to get the trade at much lower market impact, the many possibilities fairly screamed themselves to researchers, echoing and amplifying the old-line pairs trade screams heard by a previous generation two decades earlier.

All of this opportunity offered itself for reaping without requirement of capital commitment. The risk of proprietary trading was eliminated and the "new" business became infinitely scalable.

Morgan Stanley has competitors, of course. Algorithmic trading tools have been developed and marketed by Goldman Sachs, Credit Suisse First Boston, Lehman Brothers, Bank of America, and others.

10.2 MODELING EXPECTED TRANSACTION VOLUME AND MARKET IMPACT

The place to begin is the data mine. What data is available and which of it is pertinent to answering the "How much... ?" questions. Suppose that for stock XYZ there is a history of daily transaction volume data by individual trade, for over ten years. That is 2,500 days of daily transaction material. The first thing to do is examine the cumulative trade volume by day: Every stock has a distinctive character to its pattern of trading over the day, a footprint if you like. Using a one-shoe-fits-all approach, forecasting an elephant's footprint using a generic mammal footprint may work but will suffer from needlessly large inaccuracies (noise or error variance). Worse would be to use an asp's footprint (try to describe it). You can see the problem.

The problem is easily addressed by applying a modicum of specificity in the data analysis and model building. Computers don't care how many model variants they process. You should care, however; overspecificity where it is unnecessary also leads to over-large prediction variance because a finite data resource does not yield an infinitely divisible reservoir of information. The more the data is carved into different animals, the less information there is on each. If two or more animals are essentially identical (for the purpose under investigation) the data is best pooled. Moreover, the more models one tests on data, the greater the likelihood of finding a spuriously good fit. These are well known, though often ignored, details of good applied statistical analysis.

Begin looking at the data with a view to identifying a trading day pattern in transaction volume. How to characterize it? While it is unlikely that the daily pattern ten years ago is close to the daily pattern today, it would be inadvisable to assume so. Remember that reversion patterns exploited by the original pairs trade persisted with economically exploitable frequency and magnitude for a decade and a half before technological and market developments caused a dramatic change. Examine some daily cumulative transaction volume charts from ten years ago, some from five years ago some from this year. You will notice a similar form to the graph (curve) but obvious differences—faster cumulation early in the day and again late in the day comparing recent patterns to earlier patterns. Better not simply aggregate all the data and estimate an average curve then.

Look more closely at daily patterns for the last three months. That is 60 charts. Examine a three-month set from ten years ago. You notice quite a lot of overlap in the basic shapes. But look at the scales: The stock trades at much higher volumes now than it did a decade ago. Hmmm. Rescale the graphs to show cumulative percent of daily total volume. Now all graphs are on the same 0–100 scale. Aha! There is much less variability in the patterns of the last quarter. So, whether a given day is relatively high or relatively low volume, a similar pattern for the trading over the day is revealed.

How do we use this insight? One goal is to represent the curve (of cumulative percentage trade volume in a day) in a way in which it will readily yield the proportion of a day's trade volume in the market at a specific time. In other words, to provide a ready answer to questions such as, How much of the volume is transacted by 2 P.M.? There are many mathematical functions that have the generic S shape required: Cumulative density functions of probability distributions provide a natural set since distributions are precisely what are being examined here. A convenient form for statistical model building (which we have not yet considered) is the logistic function.

Pick a function. Fit it to the data. You can now readily make sensibly quantified stock-specific responses to the question: How much of the day's volume is transacted by 2 P.M.? *On average ...*

Now today happens to be a reasonably heavy trading day for the stock, with 4 million shares traded by 11:30 A.M. How many shares are expected to trade by 2 P.M.? From the estimated pattern, fully 30 percent of the day's volume is typically transacted by 11:30 A.M., and 40 percent by 2 P.M. Easily you compute 1.3 million shares are expected to trade over the next 90 minutes. You want to trade 100,000 shares. Should not have to pay much to achieve that. Right?

The foregoing analysis considered only transaction volume; price information in the record has not yet been examined. Let's redress that directly. In the set of 60 days of trading data for XYZ, there are many individual buy and sell transactions for order sizes as small as 100 shares to as large as 100,000 shares. The fill information for all orders is also recorded. Plotting order size against the change in price from the order price (or market price at time of order) and the average fill price shows a definite relationship (and a lot of variation). Once again, some of the variation magically disappears when each day is scaled according to that day's overall volume in the stock. Orders, up to a threshold labeled "visibility threshold," have less impact on large-volume days.

Fitting a mathematical curve or statistical model to the order size–market impact data yields a tool for answering the question: How much will I have to pay to buy 10,000 shares of XYZ? Note that buy and sell responses may be different and may be dependent on whether the stock is moving up or down that day. Breaking down the raw (60-day) data set and analyzing up days and down days separately will illuminate that issue. More formally, one could define an encompassing statistical model including an indicator variable for up or down day and test the significance of the estimated coefficient. Given the dubious degree to which one could reasonably determine independence and other conditions necessary for the validity of such statistical tests (without a considerable amount of work) one will be better off building prediction models for the combined data and for the up/down days separately and comparing predictions. Are the prediction differences of *practical* significance? What *are* the differences?

One drawback of fitting separate models to the distinct data categories is that interaction effects (between volume, up/down day, buy/sell, etc.) cannot be estimated. If one is looking for understanding, this is a serious omission as interactions reveal subtleties of relationships often not even dimly suggested by one-factor-at-a-time analysis. If one is looking for a decent prediction, the omission is intellectually serious (if there are interactions) but practically (depending on the nature of the interactions) of less import.

Time of day is also significant in market impact estimation—recall the analysis of the cumulative trading volume pattern over the day. Filling an order during the "slow" or more thinly traded part of the day requires either more patience for a given slippage limit or a willingness to increase that limit. Time of day was not addressed in the order size–market impact analysis outlined previously. Obviously it can be, and the obvious approach is to slice the data into buckets for the slow and not slow parts of the day (or simply do it by, say, half-hour segments) and estimate individual models for each. While the statistical modeling and analysis can be made more sophisticated, the simple bucketing procedure posited here serves to exemplify the opportunity and the approach. (Examples of fruitful sophistication include formally modeling parameters across time slices with a smooth function, and employing classification procedures such as regression trees to identify natural groupings.)

10.3 DYNAMIC UPDATING

Examining the basic patterns of daily trading volume from ten years ago and more recently has prompted the realization that patterns have changed. Immediately one is confronted by the problem of how to manage the change in predictive models estimated from the data. The first action was to use only recent data to build the model to use now. We'll assume recent time at 60 days. Now one is confronted by the question, When should the models be revised? We are once again faced with the questions about types of change, rates of evolution, and methods of dynamic updating that were discussed with respect to the reversion models in Chapter 2. The basic issues here are no different. One might reasonably elect to use a rolling 60-day window, reestimating modeled relationships each day. One might also routinely compare the latest daily pattern with the distribution of patterns seen (a) recently or (b) further distant in time to make a judgment about whether today is unusual. If it is, perhaps it would be wise to apply a "conservatism filter" to the forecasts? A measure of the rate of change could be devised (there are standard ways of comparing probability distributions, from summary statistics, including moments, to integrated measures of information), and employed to build a general dynamic updating scheme that is more flexible than the simple 60-day moving history window.

10.4 MORE BLACK BOXES

We have deliberately singled out Morgan Stanley at the beginning of the chapter because of the link to the genesis of our major theme: statistical arbitrage. But Morgan Stanley is not the only firm to have analyzed transaction data and offered tools to the marketplace encapsulating trading intelligence discovered therefrom. Goldman Sachs' operations on the floor of the NYSE—the Spear, Leeds & Kellog specialists bought in 2000—represent a gold mine potentially even more valuable than Morgan Stanley's database. Bank of America bought the technology of hedge fund Vector in 2002: "... computer algorithms will factor in a particular stock's trading characteristics *and BofA's own position in it* then generate buy and sell quotes" (*Institutional Investor*, June 2004; italics added for emphasis). Credit Suisse First Boston (CSFB) hired a former employee of the renowned and technologically advanced hedge fund D.E. Shaw, and built a tool that "processes fully 40% of its [CSFB's] order flow" (*Institutional Investor*, June 2004); Lehman Brothers and more than a dozen others are also in the business.

In addition to the developments just listed, at least one new brokerage, Miletus, has been spun out of a billion dollar hedge fund to monetize the value in the trading algorithms developed for the hedge fund's own trading. In another technology driven development, beginning with Goldman Sachs in late 2006, at least two offerings of general hedge fund replication by algorithmic means have been brought to market. As these instruments gain popularity there are likely to be new systematic pattern generating forces added to the market.

10.5 MARKET DEFLATION

Figure 10.1 depicts the market for buying and selling stocks, a generic market where buyers and sellers come together to agree on a price for mutually acceptable exchange of ownership. There are many buyers and many sellers. Lots of individual excitors. Many points of agreement. Substantial volatility.

Figure 10.2 depicts the arriving market for buying and selling stocks. The many individual buyers and sellers come together by

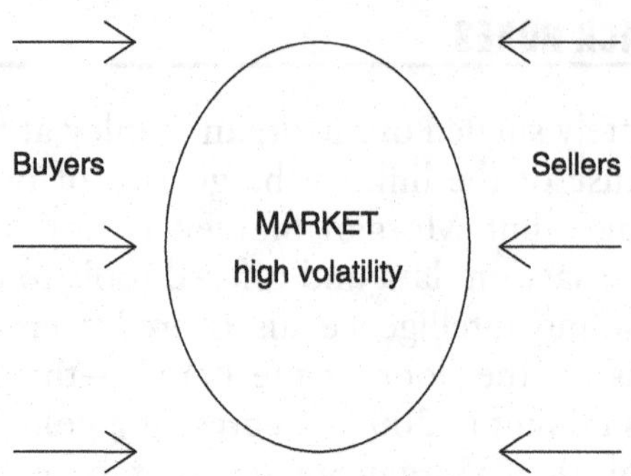

FIGURE 10.1 The way the market was

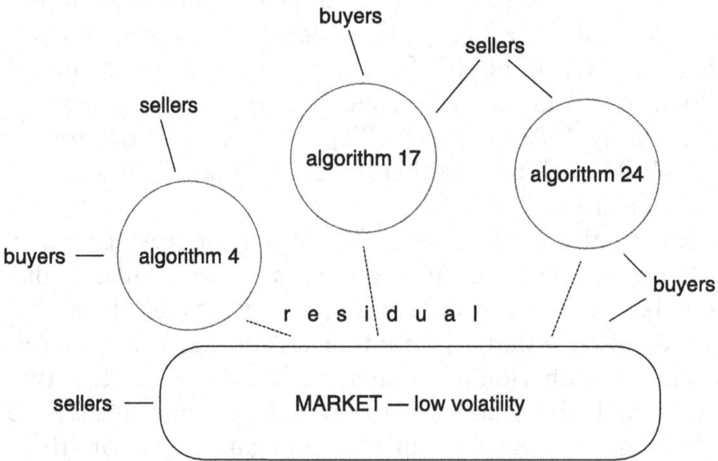

FIGURE 10.2 A deflated market model

the intermediating management of a handful of computer algorithms which internally cross a substantial portion of orders and satisfy the residual by restrained, unexcitable exchange in the central market. There are many buyers and sellers. Many points of agreement. But less unmitigated agitation than the traditional bazaar. Constrained volatility.

CHAPTER 11

Statistical Arbitrage Rising

...to worry about everything is unnerving. It is also counterproductive, for it can result in continual tinkering with a correctly operating system in response to imagined phantoms in the data.

—*Statistical Control by Monitoring and Feedback Adjustment*, Box and Luceno

By the end of 2004, statistical arbitrage practitioners had been beleaguered for a year. Investors and commentators cite performance volatility but no return set against market advance (in 2003); adduce accusative assertions of irreversible decline from visible market changes; and largely turn deaf ears to the necessary complexity of the reality (see Chapters 9 and 10 for a full exegesis).

Set against that siege is the present discourse and a return of performance since 2006. Chapters 2 to 8 set out the nature and extent of traditional statistical arbitrage opportunities, approaches to formal modeling and systematic exploitation of those opportunities, the nature of market dynamics that wreaks havoc on portfolios built and managed according to statistical arbitrage models. Chapter 9 examines one-liner condemnations of the discipline, the logic of which is: This change eliminates a part of statistical arbitrage return; the change is permanent; your opportunity set is therefore gone. The claims are found pertinent but inadequate to explain the record. The far more complex reality is no less devastating but, upon deeper reflection, cannot support condemnation. In their complexity, the enduring elements are not wholly destructive of statistical arbitrage. To the contrary, some of the more far-reaching market structural changes, set out in Chapter 10, necessarily create conditions for a new statistical arbitrage paradigm. That emerging paradigm, its driving forces and consequent statistically describable and therefore exploitable stock price patterns, is set out in this chapter. It both concludes the present volume and sets the scene for a subsequent history to be written some years hence.

A few statistical arbitrage practitioners with long and outstanding performance pedigrees continued to deliver reasonable to good returns while most have failed as described in earlier chapters. This evidence supports the claims of (a) incomplete destruction of traditional statistical arbitrage opportunities and (b) genesis and development of new opportunities, though only proprietary information could reveal to what extent the evidence supports each claim individually. Evidence in the analysis of the public record of stock price history strongly suggests that the opportunity for extracting outsize returns from high-frequency trading—intraday—is huge. From the discussion throughout this book, it is clear that exploiting that opportunity requires different models than the traditional mean reversion type. Some such models are described later in this chapter.

Patterns of stock price movements within the trading day show not reversion but momentum. There are also patterns of reversion within the day but these patterns seem to be difficult to predict (though there are claims for success here); they occur spasmodically for broad portfolios, with precursor signals that are not easily identified. Indeed it may be inappropriate to label the movement as reversion; reversal may be more indicative of the dynamic. The distinction is crucial. A reverting process assumes an underlying equilibrium to which price (or relative prices) tends to return following a disturbance away from it (the popcorn process). Equilibrating forces can be identified. The trends and reversals process makes no such underlying assumption of equilibrium; rather, the process is one that moves for more or less extended periods in one direction and then in the other without a strong link from the one move to the next (a memoryless switching process). Critical to the description and assessment are the duration and magnitude of the directional moves: They endure sufficiently long (think of sufficiently many time steps where each step is visible to a person) and the move is large enough to be exploited systematically, given necessary lags for turning point identification.

Crucial to successful modeling is an understanding of the forces in the market driving the trend creation. Penny moves are one important factor, having already removed price friction eliminating historically innate initial resistance to repeated (and therefore cumulatively large) moves. Compounding this is the increasing removal of human specialists from the price setting process as more trades are crossed automatically on electronic exchanges and by the brokerage houses' trading programs described in Chapter 10. Most significant are those "intelligent" trading engines and the significant proportion of transactions preempted as VWAP or TWAP. Old-line technical analysis may, curiously, retain some efficacy in exploiting the new intraday trend patterns; but greatest success will inhere to those whose modeling incorporates knowledge of the underlying motive forces and algorithmic trading tactics.

Far removed from underlying equilibrating forces, driven by people making judgments of fair valuation of company prospects both short- and long-term, the new paradigm is one of unemotional—uninterested—rule-based systems continually probing other similar entities. The process is mechanistic as in a geological process, water finding the lowest level. Here, however, the rules are defined by human modelers and not the laws of the physical universe, and they are changeable. Noise is omnipresent as human traders still directly preempt a sizable chunk of market activity and originate all transactions. Notwithstanding the noise, the new forces for equilibrium are searching not for fair relative prices but fair (mutually accepted by participating entities) market clearing. This new paradigm may be a reversion (!!!) to an age-old paradigm of economics: perfect competition. Now, on that train of thought one might conjure ideas of dynamic cobweb algorithms, game theoretic strategies, and perhaps a necessary repositioning of research into behavioral finance.

Volatility will remain consumed by the algorithms. Instead of human-to-human interaction either face-to-face on the floor of the NYSE or face-to-screen-to-face in electronic marts, there will be algorithm-to-algorithm exchange. A large and growing part of the emotion surrounding trading is removed, and with that removal goes volatility. Yet in this focus on algorithms, we must not forget that people still drive the system. With trades managed by algorithms implemented on incredibly fast processing computers, what might be done by algorithms designed to go beyond passive market participation to active market determination? Possibly probing other algorithms for weakness, for opportunities to subvert naivete, or to mislead into false judgment. Warfare by another name. The attraction for certain managers and the challenge for certain programmers is irresistible.

Speculation of course, I think.

11.1 CATASTROPHE PROCESS

Since early 2004, spread motions have been observed to exhibit an asymmetric process where divergence is slow and continuous but convergence—the "reversion to the mean" of old—is fast(er), even sudden by comparison. Convergence is not necessarily "to the mean" though it is in the direction of a suitably local view of the mean. The first two characteristics contrast with those of the popcorn process, which exhibits a faster-paced departure from the norm and slower return. The third characteristic, the degree of reversion to an underlying mean, also distinguishes the two processes: In the newly emerging process, the extent of the retrenchment move is far more variable than was the case for the popcorn process.

Now we enter a definitional quagmire, so careful examination and explication at length is desirable.

Contrast the classical popcorn process with the new process using Figure 11.1. The notable features of the new process are: a slow, smooth divergence from local equilibrium; fast reversion toward that former equilibrium; *partial* reversion only (in most cases); repeated moves in quick succession delineating a substantive local trend away from the underlying equilibrium. (The latter is, as in all archetypal illustrations, depicted as a constant level. In practice, it is superimposed, on long-term trend movements—for a positive trend, turn the page counterclockwise by several degrees to view the archetype.)

The critical departure in this new "catastrophe" model is the appearance of local trends within the period classically depicted as sufficiently local to be constant. The local trend (within a trend) *must* now be depicted and formally incorporated in the analysis because it is part of the opportunity driver and is crucial to the successful exploitation of the new reversionary moves. It cannot be ignored as noise on an underlying (popcorn) process.

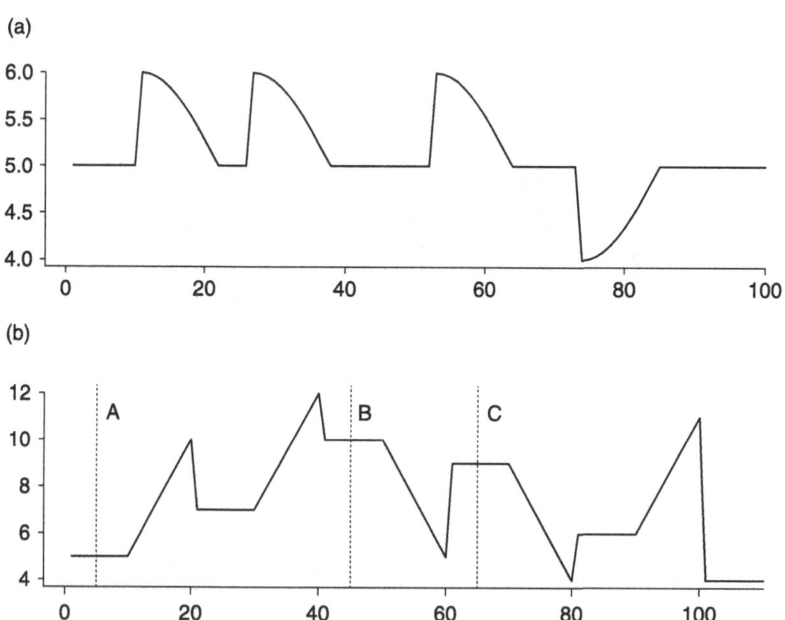

FIGURE 11.1 (a) Archetype of the popcorn process showing reversion to a mean (b) New archetype: catastrophe process

The combination of variable amounts of "reversion" and multiple moves in the same direction before a larger directional shift (singly or, again, multiple small events) is driven by the interaction of algorithmic trades. (There may be other drivers, but they remain elusive at this time.) Patient algorithms ease up when prices move repeatedly, penny by penny by penny—moves that specialists are keen on following the change to decimalization and which are undoubtedly programmed into some algorithms. What used to be a certain inertia to moves when tick size was substantive, an eighth, is now eagerness to repeatedly penny. Pennying was ridiculously lucrative at first when human traders still dominated order flow. The patience and discipline of algorithms having replaced direct trader involvement have altered the dynamics of the interaction. The results, which seem now to be clear, are the catastrophe moves we have described.

Notice the implications for the popcorn process model applied to the catastrophe process relative price evolution: zero return.

A natural question to ask is, What happens over a longer timescale than that encompassed in A–C in Figure 11.1? The description just

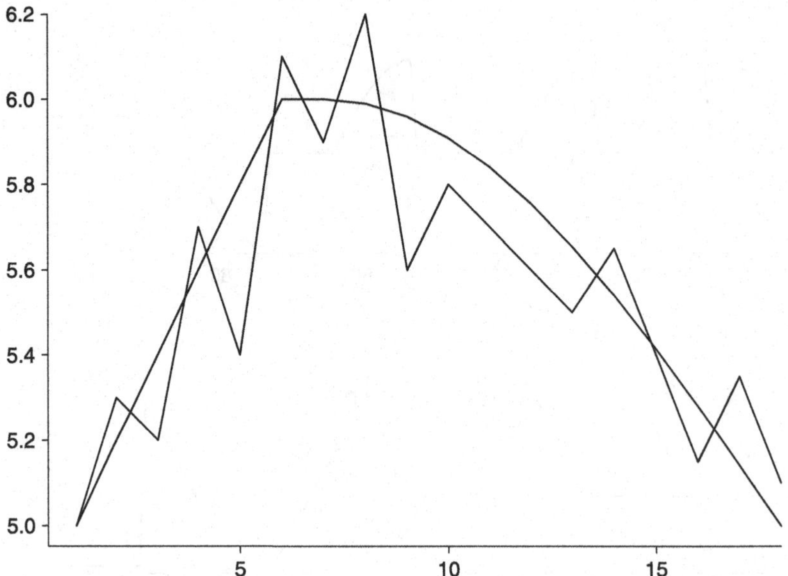

FIGURE 11.2 Extended popcorn move with catastrophe moves in detail

given, serial moves in one direction punctuated by partial retrenchment, then essentially the same in the opposite direction, as shown in Figure 11.2, and variant in Figure 11.3, sounds like little more than a sharper focus on the popcorn process, as if one simply turned up the magnification to see more of the uninteresting, picture clouding, uneconomic detail. The proper interpretation is to extend the time-scale so that the micro moves on the popcorn process become as time significant as the original popcorn move itself. Thus, the popcorn move may require as many as six (or even more) catastrophe moves to complete—a long time in a dynamic market. Popcorn's return even under these ideal conditions is reduced many fold. But the true picture, returning to Figure 11.1, is seriously worse. Over the long time period just suggested, the local mean shifts more than can be assumed away, invalidating the basic popcorn process. At best, the outcomes become distributed across a range in which the uninteresting values detract greatly from the interesting, as shown in Figure 11.4, converting an exploitable structure to a textbook or journal curiosity. Over the extended duration, the bet becomes a fundamentally dominated play; for the statistical popcorn process

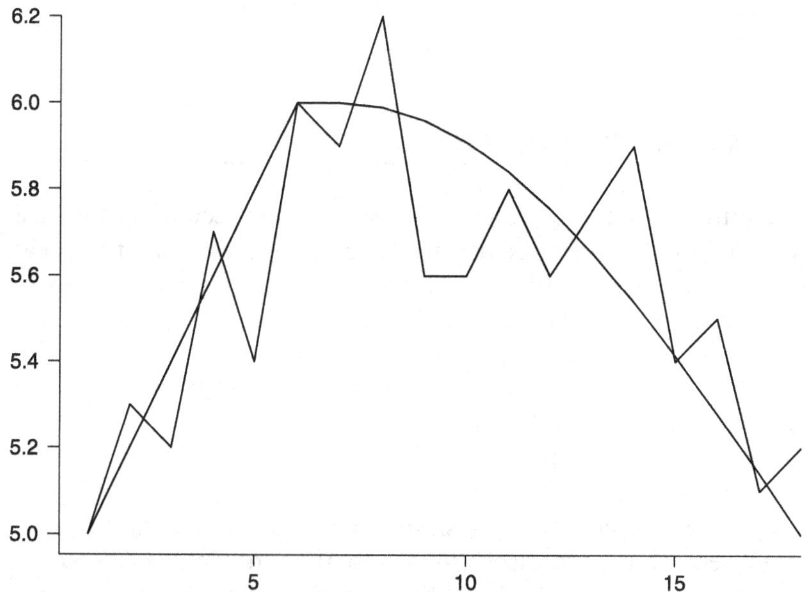

FIGURE 11.3 Variant on extended popcorn move

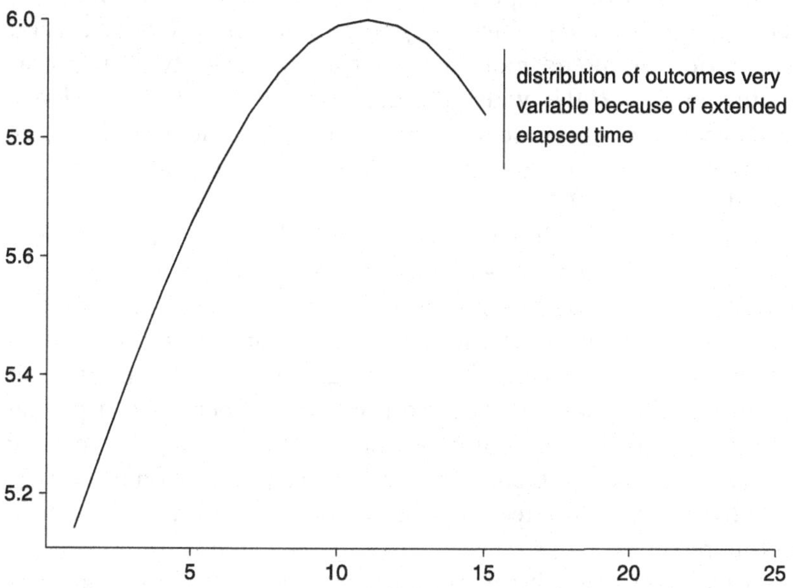

FIGURE 11.4 Extended popcorn move has variable result

not predicated on fundamental analysis, forecast power evaporates and the return along with it.

11.2 CATASTROPHIC FORECASTS

The magnitude of a catastrophe reversion is not accurately forecast in comparison to forecasts of the popcorn process. But the great variation in results from both systems, popcorn applied to popcorn data, catastrophe to catastrophe data—and where is the cutoff? Say pre-2002 for popcorn, post–mid-2004 for catastrophe, with the intermediate 18 months dominated by disruptive influences of change—means that for a large collection of bets, the statistical measure R^2 is similar. The significance of that observation for trading is an overall expectation of similar rates of return if the number of bets in a reasonable period is similar and the overall variation in the two sets of bets is also similar. Reality, of course, is not so obligingly straightforward. As the catastrophe process has come to characterize spread motions more accurately than the popcorn process, general levels of spread volatility have been decreasing (see Chapter 9). Before 2003, when the popcorn process provided a valid representation of spread motions, volatility was nearly double that prevailing in late 2004, when the catastrophe process provided a more accurate model. These outcomes are not coincidental. Both are driven by the increasing market penetration of trading algorithms (as described in Chapter 10).

A reduction in overall variance of which a similar fraction is captured by model forecasts—on the face of it, that is a recipe for a reduction in return commensurate with the variance shrinkage. But the face, too, is altered, in the shape of shorter duration moves and a greater frequency of moves. The resulting increase in the number of bets counters the lower revenue from individual bets. It is a partial counter only and is itself countered in turn by transaction costs of the increased bet count. Continued downward pressure on brokerage and trading technology fees has been and will continue to be an inevitable result.

At this point the critical question to answer is, How can systematic exploitation, trading the catastrophe signals, yield desirable economic results?

Ideally, one would like to identify the beginning of a catastrophe jump just before it occurs, allowing sufficient time to make a bet without market impact, and identify the end of the move soon after it is over to allow maximal capture of the catastrophe. Neither identification task has proven easy thus far, but approximations based on duration measures have been established.

Return to the growth and drop (or decline and jump, if you prefer the antithetical reversion) archetype catastrophe shown in Figure 11.5. Focusing on the build-up to the catastrophic move, one can identify a duration rule that signals a bet entry *k* periods following the start of the trend development. That trend onset becomes known only several periods into the move. Statistical analysis reveals a distribution of trend durations preceding a catastrophic retrenchment, and bet entry is signaled at a fixed point of that distribution. The eightieth percentile is a good operating rule.

Timely identification of the discontinuity, the change from divergence to reversion, is critical to successful exploitation of catastrophe moves. There is much less statistical forgiveness in the timing of a bet entry than was the case for popcorn moves. The relative speed of

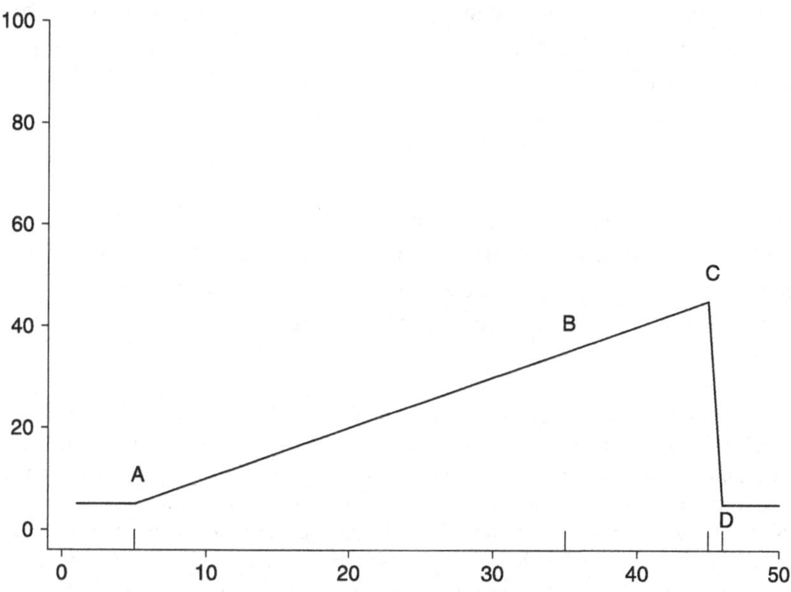

FIGURE 11.5 Catastrophe move archetype

the catastrophe reversion makes the opportunity loss from late identification of a catastrophe much greater than for late identification of a popcorn move. Failure to enter before the cliff edge, point C in Figure 11.5, essentially means missing the full opportunity. The popcorn move seen in Figure 11.3 is quite different. Late entry will lower return on a bet, but only marginally. Modeling and trading catastrophe moves must embody a higher state of alertness.

11.3 TREND CHANGE IDENTIFICATION

There is a rich statistical literature on change point identification with many interesting models and approaches providing plenty of fascination. Our purpose here is mundane by comparison, though challenging nonetheless. (If any kind of pattern recognition in financial data were not so challenging, we would hardly be writing and reading about it.) An extremely useful approach from statistical process control relies on Cuscore statistics (Box and Luceno 1987).

Consider first a catastrophe superimposed on an underlying rising series. Figure 11.6 shows a base trend with slope coefficient 1.0 with a catastrophe move having slope coefficient 1.3 beginning at time 10. Let's see how a Cuscore statistic for trend change works in this readily understood illustration. The Cuscore statistic for detecting a change in trend is:

$$Q = \sum (y_t - \beta t)t$$

where y_t is the series of observations, β is the regular slope coefficient (the rate of change in the observation series per unit time[1]) and t is a time index. The Cuscore is shown in the lower panel of Figure 11.6. Despite having seen many such graphs for many kinds of time series,

[1]Models for parametric change have much wider applicability than just time indexed series, which is our focus here. Spatial models, where observation series are indexed by geographic location rather than sequentially in time, are employed in many sciences from geology to seismology (which often has both time and space indexing) to biology (EEG readings form specific patterns across the head as well as particular temporal development at each site). In stock price analysis, indexing by trade volume is employed in trading algorithms (see Chapter 10) and by some statistical arbitrageurs.

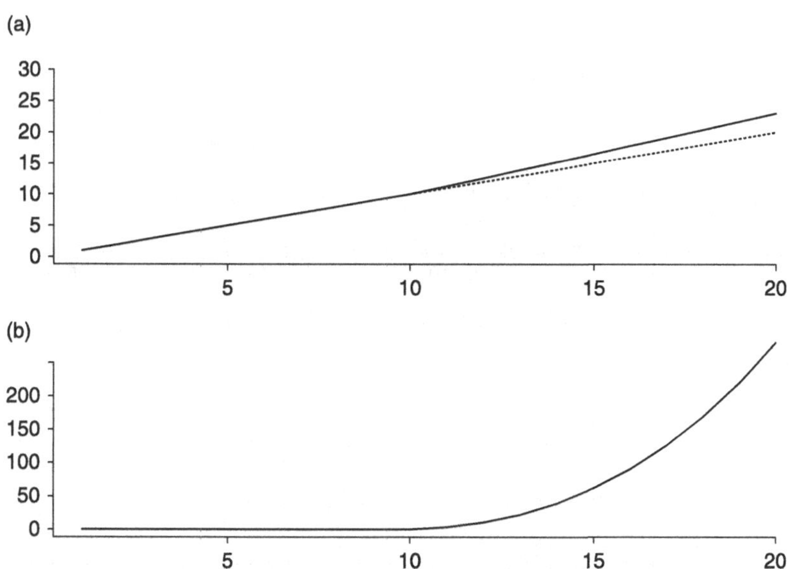

FIGURE 11.6 Identification of trend change: (a) gradient 1.0 shifts to 1.3 at time 11; (b) Cuscore

I never cease to be amazed at the seemingly magical way in which the detection statistic uncovers and displays the incontrovertible evidence of change. The slope increase of 30 percent from initial value 1.0 to subsequent value 1.3 looks, as just written, substantial. Thirty percent is nearly one-third and that surely is substantive and ought to make us take notice. But the graph generates a very different perception. Were it not for the dashed continuation line, we would be hard-pressed to notice the kink in the line at time 10. The visual discordance is too small. Pictures may paint many words but here is a case in which the words are more dramatic.

The dramatic shift from constant to exponential increase in the Cuscore statistic recovers the situation efficiently and effectively. Now, how does the Cuscore perform when observed values do not fall neatly on prescribed mathematical lines? Figure 11.7 adds random noise (Student t distribution on five degrees of freedom, a heavier tailed distribution than the normal) to the series depicted in Figure 11.6. If the slope increase was visually difficult to discern previously, it is practically impossible now. How does the Cuscore statistic fare?

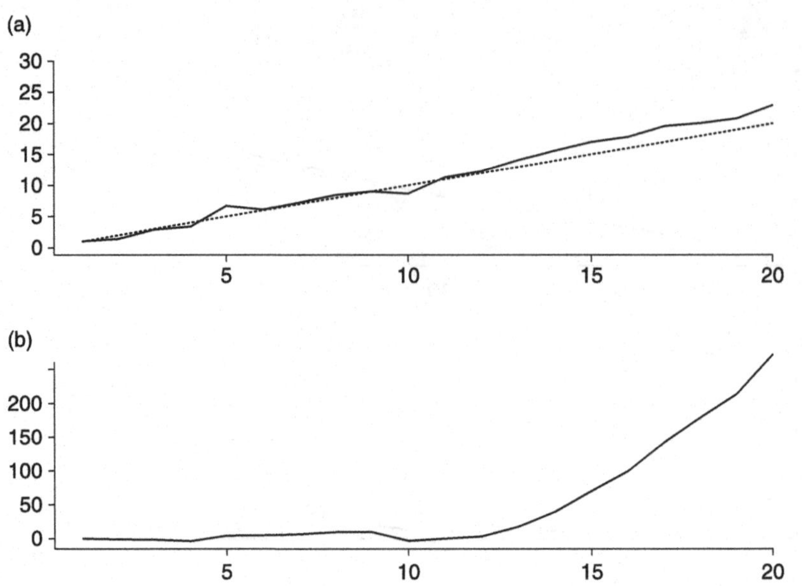

FIGURE 11.7 Cuscore identification of trend change for noisy data: (a) time series; (b) Cuscore

The illustration in Figure 11.7 is more dramatic than that in the previous figure over which we were so excited. Eyeball analysis is of little use here, serving only to generate eyestrain. The Cuscore statistic, by stark contrast, signals a high probability of a trend increase by time 15 and practical certainty one or two periods later.

11.3.1 Using the Cuscore to Identify a Catastrophe

In the foregoing examples, the underlying trend was constant and the coefficient β in the Cuscore statistic was set to the known value of 1.0. Unfortunately, financial series do not come to us packaged with a convenient quantification of underlying rate of change. We have to work with the raw observations. Completing the task of identifying a catastrophe move in a spread series requires the specification of the underlying trend prior to the potential catastrophe. At first thought, one might suggest using a local average computed using an EWMA as recommended in Chapter 3. But a chicken and egg difficulty becomes apparent almost as soon as the suggestion is made. The local average, be it EWMA or some other formulation, will be contaminated as soon as the catastrophe move begins. The Cuscore will be in the impossible position of detecting a change in slope using not the new and old slope quantification but the new quantification and itself. What is needed is an estimate of the trend if no change had occurred allowing that a change *may* have occurred. Since the timing of a potential change is unknown, what can one do?

Two simple strategies have some efficacy. For underlying trend estimation in the presence of potential catastrophes, one can use a substantial period of time, such as several multiples of the catastrophe duration, obtained from inspection of the empirical distribution of catastrophe moves in tradeable spreads.[2] A second scheme is to employ an estimate of the slope coefficient obtained from the EWMA ordinarily found to be sensible for the series under study. The formula for the modified Cuscore statistic becomes:

$$Q = \sum (y_t - \hat{\beta}_t t)t$$

where $\hat{\beta}_t$ is the estimated current slope coefficient. Derivation of $\hat{\beta}_t$ is given in Appendix 11.1 where the Cuscore statistic for trend change detection in stock prices is examined in some detail.

Operationally, the statistic works well, but there may be superior detection procedures for early catastrophe move identification. One possibility (explored in Appendix 11.1) is to employ a lagged local trend estimate to avoid the chicken and egg problem. Since it is "known" that catastrophe moves are identified five time-steps after onset, it is reasonable to estimate the underlying trend by excluding at least the five most recent series observations.

Why not lag the EWMA by more than five observations, just to be "safe"? (Technically, increase the probability of detecting a catastrophe move in the presence of noise when the catastrophe build-up is "gentle"). That is a question of the modeler's art as well as detector performance characteristics. Significant relevant considerations

[2]Identifying catastrophe moves in past data is far simpler than doing so on line. Any candidate move can be confirmed by later data before it is assigned a classification and employed in study of identification and characterization rules. On-line decisions must be made, and trading decisions taken, before such confirmation is possible. Of course, confirmation is eventually possible but that is after one has made a profit or incurred a loss.

as you pursue this investigation are:

- What is the distribution of differences between underlying trend and catastrophe precursor trend?
- What is the distribution of the magnitude of catastrophic reversions?
- What is the relationship of the magnitude of catastrophic reversions to the duration of the build-up and the magnitude of the difference between underlying trend and catastrophe precursor trend?
- What is the set of catastrophes that is economically desirable to capture?
- What is the cost of a false identification of a catastrophe?

Good hunting!

11.3.2 Is It Over?

A popcorn move finishes when the spread series returns to the (local) mean, plus a bit beyond the mean contributed by stochastic resonance. When is a catastrophe move complete? I have most proficiently answered this to date through a fixed duration following detection of a spike in the opposite direction of the development of the catastrophe. If the catastrophe developed as an increase over the underlying trend, as in the previous examples, then the catastrophic change ending the move would be a sudden decrease.

I have not answered the question with the success achieved in other areas of statistical arbitrage modeling. Whether the catastrophic move is a single large move or a trend over several periods, the onset is revealed by Cuscore monitoring. A second modified Cuscore statistic is employed, recognizing the nature of the move: The underlying trend is now the build-up of the catastrophe itself so the appropriate estimate is based on the interval starting at the estimated onset of the catastrophe and ending one or two periods before the latest observation. The Cuscore is specifically looking for a spike in the opposite direction of the catastrophe build-up; here we allow a spike to be over one or two periods, hence the need to exclude the latest couple of observations from the trend estimate. Including them would put the Cuscore in an impossible position similar to that described earlier for the catastrophe onset detection.

Best efforts to date in specifying a bet exit rule, catastrophe over, is a combination of the duration and magnitude of the catastrophic move. A significant danger is waiting too long and getting caught in a subsequent catastrophe which negates the gains of the first. There is plenty of room here for improvement in modeling and trade rule prescription.

11.4 CATASTROPHE THEORETIC INTERPRETATION

Algorithms are formulaic and primitive; there is no comparison to human consciousness. Most traders are inconsistent and unfaithful to their model(s). Algorithms are dumbly consistent, unimaginative. Still, with many algorithmic interactions taking place in the market there may be emergent behaviors, unpredictable from pure analysis of individual algorithm to algorithm interaction.

Examine the Catastrophe surface[3] shown in Figure 11.8. The catastrophe move, a slow build-up then a sudden drop, is created by

---

[3]I chose to call the new reversion pattern the catastrophe process, rather than popcorn 2 or some other label, because it is catchy and does capture the rather different move dynamic than is both suggested by the name and exhibited by the popcorn process. The development of an explanatory model of investor behavior, which might represent why the new style moves occur, is separated from the description and exploitation of those moves. The underlying elements of algorithm-to-algorithm interaction and growing popularity and use of trading algorithms in place of direct human action are undisputed. They are observed facts. Stock price histories are also incontrovertible facts. The patterns I have discerned in those histories are debatable: There is a lot of noise in any example I could show.

Trading models built to exploit the dynamics represented by the popcorn and catastrophe processes have undeniable track records. That is existential proof of model efficacy and supports the validity of the pattern descriptions, but it does not prove any theory of why the patterns are the way they are. The popcorn process has been so long established and so widely exploited at multiple frequencies that providing a rationale has not received much attention. The rise of a new pattern with the background of failure (in terms of economic exploitation) of the old also does not *require* a rationalization. If it persists and statistical arbitrageurs begin to discover it and churn out decent returns, once again investors will experience their own catastrophic shift from skepticism (fear of loss) to hope (greed).

While a rationalization is not necessary for the rise of the phenomenon of reversion by catastrophe, an understanding of market forces driving new dynamics and a cogent, plausible theory of how those forces interact and might produce

(*Continued*)

continuous moves through a two-dimensional space. The dimensions correspond to a "normal" factor and a "splitting" factor in catastrophe theory parlance. At low levels of the splitting factor, variation in the normal factor causes smooth variation in the outcome surface. At high levels of the splitting factor, movement in the normal factor generates outcomes in two distinct regions, separated by a discontinuity—the catastrophic jump. The discontinuity is asymmetric: Jumps "up" and jumps "down" occur at different levels of the normal factor for a constant level of splitting factor; this is known as hysteresis, commonly interpreted as inertia or resistance. (Figure 11.9 shows a cross-section of the catastrophe surface, parallel to the normal axis, at a high level of splitting factor, illustrating the asymmetric jump process.)

This is the classical description of the two-dimensional cusp catastrophe. Application to stock price development identifies "avarice" with the normal factor and "fear" with the splitting factor. Consider a movement over the surface beginning at A, with a low level of fear.

---

emergent patterns is necessary to promote unbiased critical attention in the formative period. The simple catastrophe theory model presented in the text is offered as one possible way in which identified market forces recently introduced and growing in influence as old behaviors and interactions are supplanted might be understood. The catastrophe model is a plausible representation of what is currently known, but it is not a formal model from which predictions can be made. V. I. Arnold in *Catastrophe Theory* acidly remarks that "articles on catastrophe theory are distinguished by a sharp and catastrophic lowering of the level of demands of rigor and also of novelty of published results." You have been cautioned.

Arnold further remarks, "In the majority of serious applications... the result was known before the advent of catastrophe theory." The strong implication in our context, despite the lapse of 20 years since Arnold wrote, is that even if the representation and interpretation of the model presented is valid, it is probably better (more rigorously, more convincingly) constructed using tools other than catastrophe theory. I am, in fact, engaged in research using game theoretic tools to model trading algorithm interactions. This work is at too early a stage of development to report here. Finally, to quote Arnold again, "In applications to the theory of the behavior of stock market players, like the original premises, so the conclusions are more of heuristic significance only." My premises are rather more than heuristic, algorithm-to-algorithm interaction and increasing dominance of algorithms and removal of direct human interaction, and patterns discerned from stock price data history being there for anyone to inquire of. Nonetheless, it is quite right to regard the catastrophe model of market agent behavior as heuristic. In keeping with Arnold's tone, I propose to describe the model as the *Tadpole theorem* with the explicit intention that it is just a little bit of Pole!

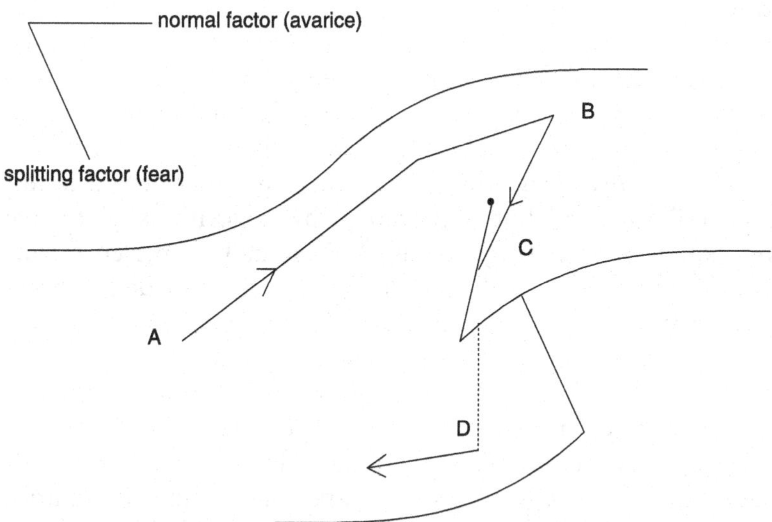

FIGURE 11.8 Catastrophe surface

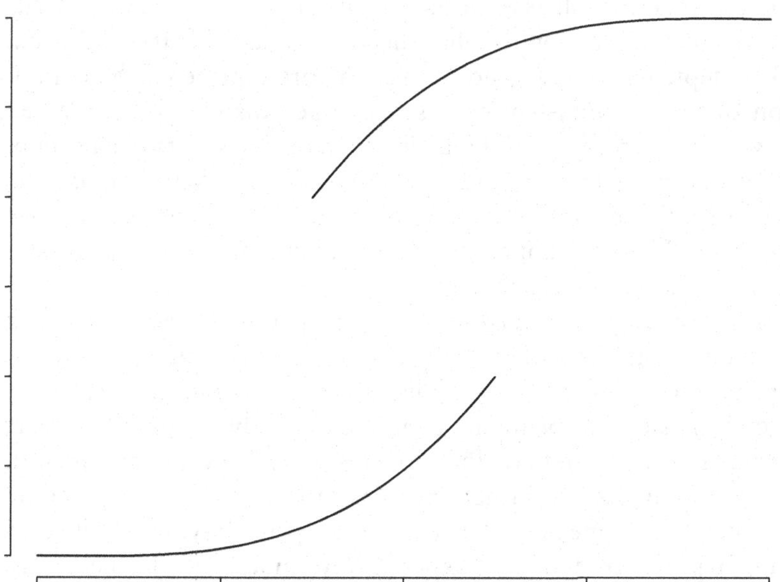

FIGURE 11.9 Cross-section of catastrophe surface at high level of splitting factor

The price develops smoothly in the direction of B with increasing avarice. As the price increases further to C (the surface is tilted upward along the splitting factor axis) fear begins to infect participants. Eventually, fear surpasses avarice as the dominant concern and there is a quick price pullback. What is the nature of the fear? Simply that the divergence in price from recent local trend is not fundamentally justified but is promoted by (algorithms') attempts to over-exploit buyers. (Algorithms don't actually experience fear, or have any experience at all, nor do they act from or demonstrate emotion. Bear with the sloppy, informal use of descriptive language: This is a work in progress. I have not by any means established "the" theory. Indeed, as you can see, I am still working on proper explication of what I have hypothesized about the process underlying observed price dynamics.) To repeat, algorithms have no conscious experience. However, algorithms do encapsulate learning about price movement dynamics (see Chapter 10), knowledge of how much is to be given up or gained through backing away from the market, and waiting. All this as well as information on current market moves feeds into a calculated reaction that has the appearance of fear—pullback.

The depiction of fear and avarice factors represents the combination of participants—buyers, sellers, specialists—as represented through their algorithms. The avarice axis measures the maximum state of avarice affecting traders and specialists: Whoever has the greediest sentiment of the moment dominates interactions and price movements. In like manner, the fear axis measures the maximum state of fear infecting participants.

As buy pressure is seen by the specialist, pennying begins. Trading algorithms, typically with some pricing room permitted in order to complete trades, follow the specialist up. Responding, the specialist's avarice increases and pennying continues (possibly picking up pace, though the description here does not require that level of specificity). As these interactions continue, price is moved higher until trading algorithms determine that it is time to suspend buying: Calibrated on much previous data to "expect" how much will be necessary to complete trades, unemotional algorithms display saintly patience. Buy pressure eases. Immediately the specialists' avarice turns to fear. Keeping price high will generate no profit if buyers stay mute and there is no business. Sellers share the fear. Price drops precipitously (in comparison with the rise) to rekindle buyer interest.

One might ask, Why not smooth decline? Because reaction to fear is different from satisfying avarice (whether it is fear of selling too cheaply or buying too expensively), notwithstanding algorithms. Remember that algorithms are designed and coded by people. Patience. Wait for a significant decline. Therefore, without intermediate activity, downward pennying accelerates and, in many cases, is observed as a multipenny catastrophic drop.

Satisfied that patience has paid off, the cycle begins over again, very likely from a starting price higher than the starting price of the original move, as these short-term catastrophic retrenchments are usually partial. Enthusiasm, avarice, builds again quickly and price races ahead of the sustainable growth path. Realization sets in, fear, and equilibrium is quickly, if temporarily, restored.

How does this description of algorithmic interaction and the resulting behavior of stock prices relate to spreads? Directly. Stock prices move at differential rates as they always have. The catastrophe moves of individual stocks naturally combine to generate catastrophe moves in spreads. Dynamics are different; scaling is different. But the basic description is identical.

11.5 IMPLICATIONS FOR RISK MANAGEMENT

A valuable risk management tool in the successful management of many statistical arbitrage models is the so-called hurdle rate of return. A model's forecast function provides an explicit expected rate of return for any contemplated bet. Managers typically specify a minimum rate of return, the hurdle, which must be satisfied before a bet is made to avoid collections of bets that are, in the aggregate, probabilistically sure losers. In times of perceived heightened general risk, typically exemplified by increased volatility, actual or expected, a standard practice is to raise the hurdle. This prophylactic action is designed to avoid entering reversion bets early, while divergence is still a strong force, thereby avoiding initial losses and, hence, increasing return. The tactic is a broad sweep action that is appropriate when concern is of a general increase in variation not focused on specific market sectors or stocks. (The tactic can, of course, be directed toward specific market sectors or other collections of stocks if there is reason to be so concerned.)

For the popcorn process, the basic forecast function is a constant, the value at any time being reasonably computed as an EWMA (with more sophisticated modelers employing local trend components, too, depending on the time scale over which the move is exploited). When the spread pops, the expected return is calculated as a fraction of the deviation between the spread and the forecast value. When volatility is expected to increase, the pops will be expected to increase in magnitude; waiting for larger pops is obviously sensible. (Slow, rather than sudden, increases in volatility are automatically managed, feeding into dynamic recalibration of models. The scenario we are concerned with here is an increase of sufficient magnitude in a short interval that is outside the capacity of automatic model adjustment. That is a risk scenario rather than ordinary evolution dynamics.) The point is that the expectation-based information is not accessible to the model from data analysis, but it can be communicated by the modeler.

Are the considerations of risk, sudden nonspecific increases in volatility, any different from those just articulated when considering catastrophe moves? At first blush it does not appear so. Catastrophe moves are a convergence following a divergence, so rescaling for a spike in volatility is just as relevant as it is for popcorn (or other reversion) models. That first blush might be of embarrassment upon further reflection. Since early 2004 when the catastrophe process emerged as the better descriptor of local price and spread motions, the general level of market (and spread) volatility has been historically low (see Chapter 9). We do not have any empirical guidance on what will happen when volatility spikes. Rescaling of local catastrophe moves may be the result. But it could easily be something different. A good argument can be made that increased volatility will swamp the catastrophes, certainly sinking the ability to identify and exploit them on line, leading to the return of the popcorn process. Is such a development more than theoretically conceivable if the hypothesis of algorithmic interaction driving price dynamics is and remains true? What would cause volatility to spike? People, of course. Algorithms are tools. Ultimately, people drive the process. We are largely in the realm of speculation at this point. Here are a couple of further points to guide your thinking:

- Waiting longer in a local trend: a duration criterion rather than expected rate of return criterion. (Is there a return forecast that can be combined?)

- Waiting longer for a bigger build-up means fewer opportunities *and* the catastrophic response is unchanged because in the catastrophe move, the reaction is not to the mean but toward an old, not really relevant, benchmark level.

11.6 SIGN OFF

The new paradigm is as yet in an inchoate state. It is actually two paradigms, a mix of a continued variant of the old reversion paradigm as interstock volatility increases, and the new trend and reversal paradigm just outlined.

Traditional interstock volatility driven reversion plays may stage a resurgence in appeal as a source of systematic return. Rising interest rates, increasing entrepreneurial risk-taking activity, or possibly a sudden recession-induced market scramble, are the drivers of this potential. The potential is there but the extent of the opportunity will be limited, returns constrained by the structural impact of decimalization, patient institutional trading (VWAP and other algorithms), and simple competition (Chapter 9).

The promise of the new paradigm is certain. However, it is not yet screaming—perhaps this kind of scream will only be heard, like Santa's sleigh bells, by believers?

APPENDIX 11.1: UNDERSTANDING THE CUSCORE

The Cuscore statistic for detecting a change in trend was developed by statisticians working in industrial process control where the goal is to be alerted as soon as possible when a process requires adjustment. An example is production of ball bearings of a specified diameter. The target mean (diameter) is known. Samples of ball bearings are taken sequentially and measured, the average diameter calculated and plotted on a chart. Deviations from the target mean occur over time as the production machine is subject to wear. Ball bearing diameters begin to increase. A plot of the Cuscore statistic reveals the onset of wear very quickly. (In practice, the range of diameters in the samples would also be monitored; different kinds of machine wear create different kinds of output variation.)

In engineering applications, like the ball bearing example, the underlying level is a known target value. Thus, in the Cuscore for detecting a change in trend, $\sum(y_t - \beta t)t$, the slope coefficient, β, is given. As we noted in section 11.3.1, the situation with stock price data is different. There is no target price at which one can direct a causal mechanism (notwithstanding hopeful prognostications of analysts). Therefore, to detect a change in price trends, an estimate of what the trend is changing from is required. Calculating an up-to-date estimate of what the trend is believed to be, a local trend estimate, is the way forward. Monitoring a time sequence of local trend estimates itself provides direct evidence about change therein.

In this appendix we examine the detail of the Cuscore for trend change detection. The study reveals how the Cuscore works and the problems inherent in the use of locally estimated trends. This latter is crucial. Insight about how estimated trends affect the Cuscore is critical to successful implementation of the detector and, hence, to online exploitation of catastrophe moves. Without timely identification, there is no economically desirable real opportunity.

In Figure 11.10, the line ABC is an archetype of a change of trend, the first segment, AB, having slope 0.5 and the second segment, BC, having slope 1.5. The dashed line BD is the continuation of line AB. The dashed line AE is parallel to segment BC, having slope 1.5. We will use these straight line segments—suppose them to be noise-free price traces to fix ideas, if that helps—to demonstrate the effects on the Cuscore statistic from different assumptions about an underlying trend when looking for a change in that trend. Knowledge of results in the noise-free, theoretical model will guide our expectations when we investigate noisy price series.

The Cuscore, $\sum(y_t - \beta t)t$, is the cumulative sum of deviations of the observation series, y_t, and the expected value assuming the slope β. In Figure 11.10, that translates into the vertical separation of y from the line segment AD. The first observation is that all points on AD will generate a zero contribution to Q. If there is no slope change, Q is identically zero.

When the slope changes, observations depart from the expected value under the base model (of no change). Values of y along the line segment BC exceed the expected values on line segment BD by an increasing amount with time. Cumulating these deviations in Q we obtain the trace labeled 1 in Figure 11.11.

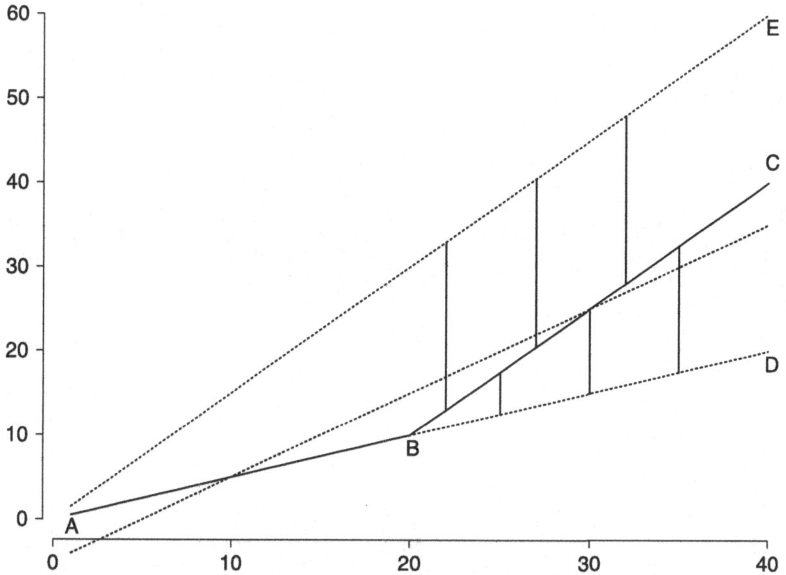

FIGURE 11.10 Trend change archetype and Cuscore contribution detail

Now suppose that we do not know in advance the slope of line segments AB and BC or that there is a change in slope at B. Suppose instead, that beginning at A, our best understanding is that the process should exhibit a slope of 1.0 as shown by line segment AC (not drawn to reduce clutter). The Cuscore is shown as the trace labeled 2 in Figure 11.11. Once again, the visual appearance of the Cuscore is startling. Deviations of a series from an assumed base model—a difference in the slope of a trend—are made starkly evident. This second example reveals both the occurrence of a change (the inflection point in the Cuscore trace) and the information that the series begins with a smaller slope than hypothesized and switches to a slope larger than hypothesized.

At this point, you probably have an inkling (or more) about the next few steps.

Review Figure 11.2. The first three catastrophe moves compound a strong positive trend; the subsequent moves compound a variably declining trend. How can we operationally, in real time, provide the Cuscore with a reasonable chance of detecting the superimposed catastrophes from the underlying, longer-term trend changes?

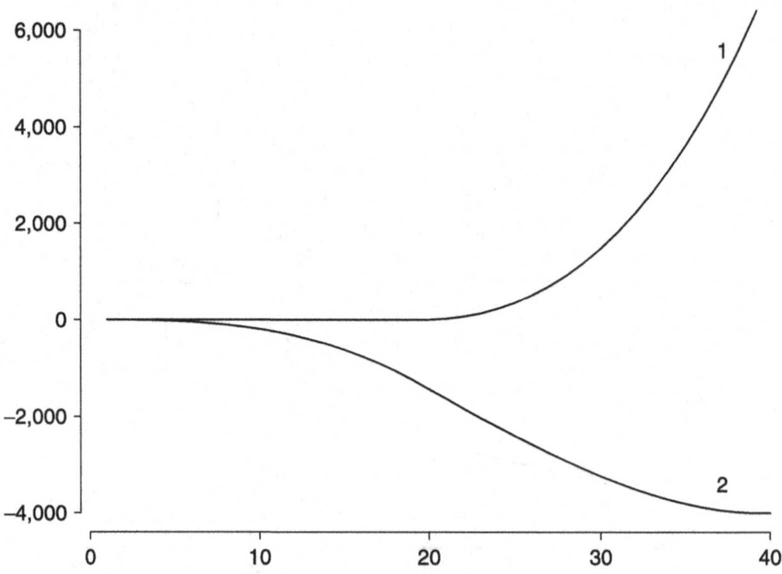

FIGURE 11.11 Cuscore with $\beta = 0$ and $\beta = 1$

An answer offered in the main text is to use a local trend estimate. Let's examine how the Cuscore behaves when a known, constant trend is replaced with a local estimate.

In Figure 11.12, the EWMA consistently underestimates the real series; that is a well known feature of moving averages, weighted or otherwise, which are not designed to project a persistent trend. The Cuscore reflects the "always trying to catch up" condition showing an increasing value from the beginning of the series. The slope change is captured, the rate of increase in the Cuscore is picking up, but the strength of inference is slow to build compared with the Cuscore using a known constant trend. The slowness problem comes directly from the use of the EWMA after the slope change. In Figure 11.12, the Cuscore contributions are the differences between the new slope and the projected old slope (vertical differences BC–BD) exemplified by $p - q$. With an estimated level, the projection of the initial trend AB to BD generating $p - q$ is replaced with the EWMA generating the much smaller deviance $p - r$. This is the chicken and egg problem. We need to project the early trend beyond the change point, which is unknown, to quickly detect that change point!

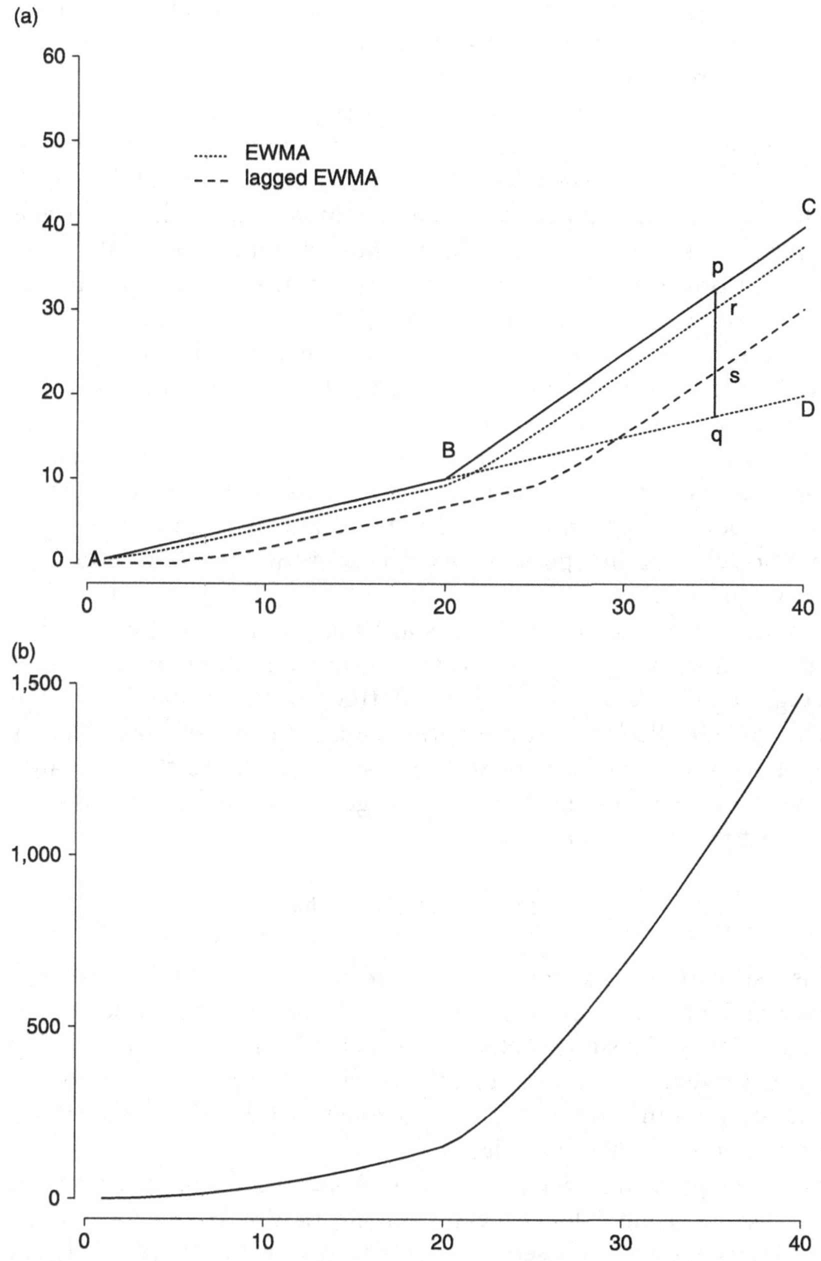

FIGURE 11.12 (a) Cuscore contributions using local mean and (b) Cuscore using local mean

Updating the local mean estimate after the slope change reduces the sensitivity of the Cuscore to detect that change. This suggests that sensitivity might be recovered by delaying the local mean update. What happens if a lagged local mean estimate is used in the Cuscore? Returning to Figure 11.12, the Cuscore contribution postchange increases from $p - r$ to $p - s$, much nearer to the desirable $p - q$. Unfortunately, this move does not eliminate the chicken and egg problem; it simply relocates the henhouse! While the postchange contributions to the Cuscore are indeed larger, so are the prechange contributions. Thus, accurately distinguishing a change in the Cuscore trace is not more easily accomplished: Premature signaling may be the frequent result. Reducing the lag—we used five periods because analysis of a catalog of identified catastrophe moves in price histories strongly suggested that most such moves with subsequent economically exploitable catastrophe retrenchments are identifiable five periods into the move—might help, but as soon as we move from noiseless archetypes to noisy real data, the situation returns to nearly hopeless.

What we are searching for is something in the data that quickly and consistently registers a substantive change following the trend change. In Figure 11.12 the EWMA trace responds quickly to the trend change. Perhaps an estimate of local trend from the EWMA might be a sensitive diagnostic? Figure 11.13 shows the estimated slope coefficient computed as the average change in the EWMA over the most recent four periods:

$$\hat{\beta}_t = 0.25(\text{EWMA}_t - \text{EWMA}_{t-4})$$

This estimate shows none of the tardiness of the EWMA-based Cuscore. Unfortunately, as soon as even modest noise is added to the original series, the slope coefficient estimate deteriorates considerably as a sensitive diagnostic, though the sensitivity is greater for a longer window when the underlying trend is constant other than at the point of focus here, as shown in Figure 11.14.

At this point we have two candidates for trend change detection, the Cuscore using a local mean estimate (EWMA) and local slope coefficient estimates based on the EWMA, each of which looks somewhat promising. The Cuscore signals the change but is tardy, the slope signals the change but is also tardy when noisy data is

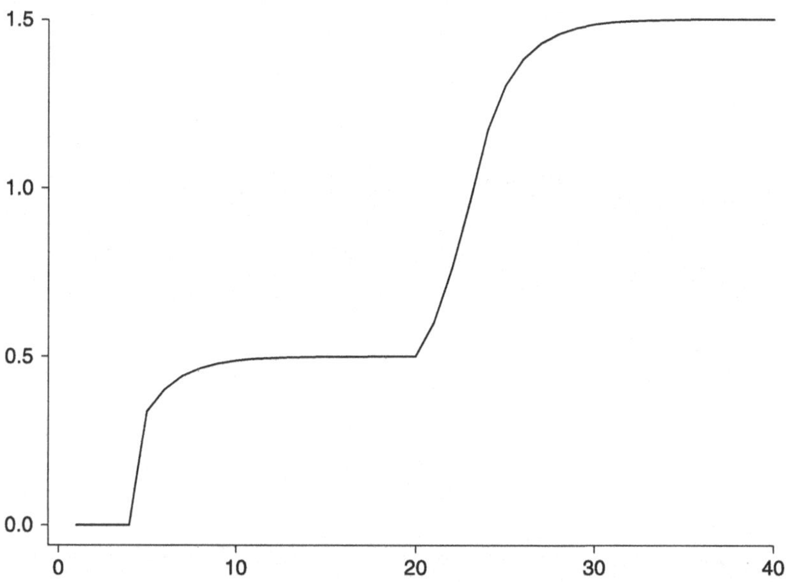

FIGURE 11.13 Estimated slope coefficient, $\hat{\beta}_t = 0.25(\text{EWMA}_t - \text{EWMA}_{t-4})$

examined. Perhaps combining the two might amplify the result? What do two "tardies" make? Before we reveal that, let's review the collection of Cuscore statistics examined so far. Figure 11.15 demonstrates the collection of Cuscore statics introduced in this appendix applied to the noiseless trend change series. Q_{theo} is the original Cuscore in which the initial trend is known. Q_{mm} is the Cuscore using a local mean estimate (EWMA), Q_{mmlag} is the Cuscore using a lagged local mean estimate, Qb_1 is the Cuscore using a locally estimated slope, $Qb1_{lag}$ is the Cuscore using a locally estimated slope from the lagged local mean, and Qb_{true} is the Cuscore using the actual before and after change slope coefficients. That's a lot of Cuscores!

Q_{theo} and Qb_{true} are theoretical benchmarks we would like an operational Cuscore to approach as closely as possible. Qb_{true} is singularly interesting. Return for a moment to Figure 11.10. Qb_{true} cumulates deviations from the known line segment AB, so the value is identically zero for $t = 1$ through $t = 20$. At $t = 21$ we switch from the old slope coefficient $\beta = 0.5$ to the new slope coefficient $\beta = 1.5$ and thence begin cumulating deviations between observations on line

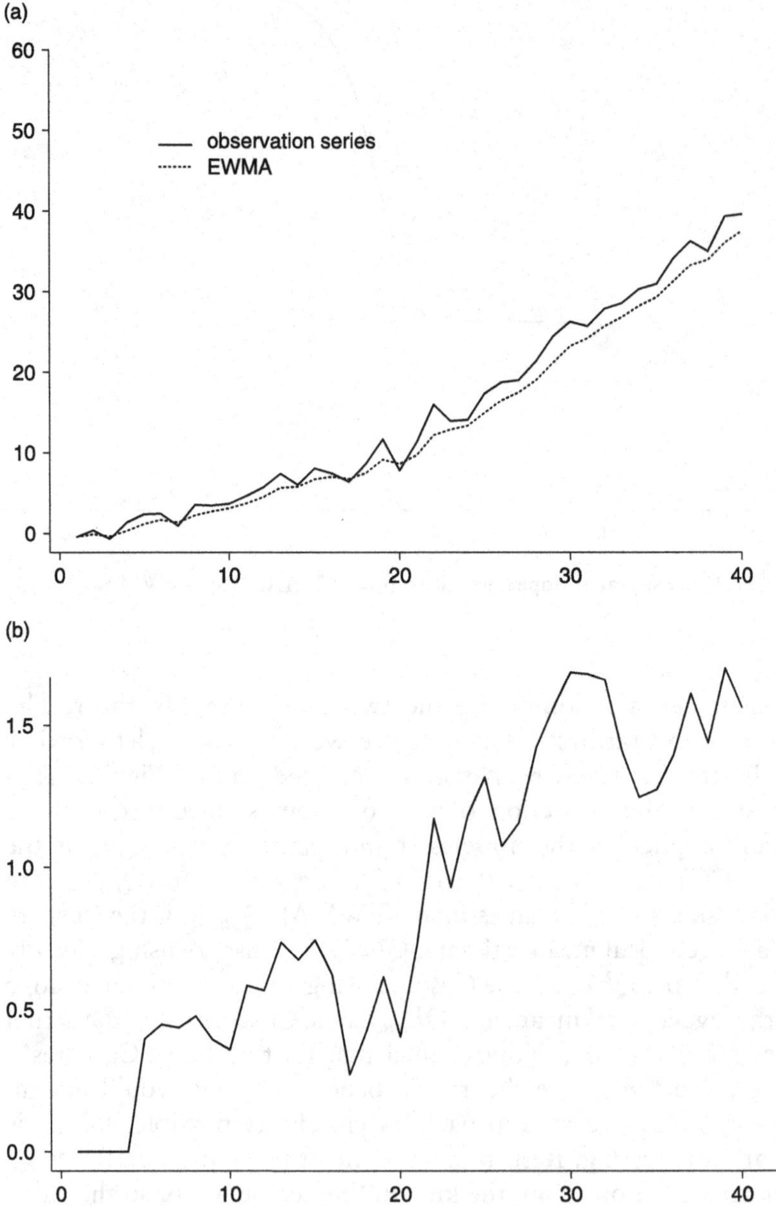

FIGURE 11.14 (a) Trend change with noise; (b) estimated slope coefficient, $\hat{\beta}_t = 0.25(\text{EWMA}_t - \text{EWMA}_{t-4})$

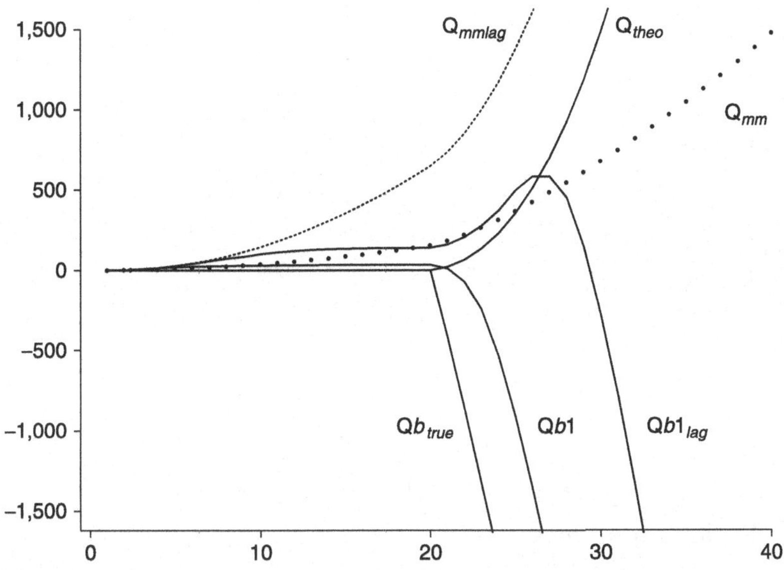

FIGURE 11.15 Cuscores for several models

segment BC and the line AE, which is the new base model, assuming gradient $\beta = 1.5$ from inception.

AE is parallel to BC (both have gradient $\beta = 1.5$) so the growth of Q is linear as the deviations are constant. This contrasts with the standard Cuscore in which the individual deviations increase sequentially (excluding noise) and, hence, the cumulative sum grows faster than linearly (Q_{theo}). Qb_{true} has the initial advantage over Q_{theo} because the deviations begin large, hence, the speed of detection of change is faster. The advantage is a function of the relative size of the two gradients and the time origin of the cumulation—the duration of segment AB. In our task of identifying catastrophes, the larger the prechange duration, the greater the discrepancies (AE–BC) feeding the Cuscore, the greater the initial advantage over the standard Cuscore and, therefore, the sooner the likely identification of a trend change. Which of the noisy sample versions of the theoretical benchmarks, $Qb1$ or Q_{mm}, dominates in practical catastrophe identification depends on the dynamics of the catastrophes and precursor periods.

Earlier we remarked that Q_{theo} and Qb_{true} are theoretical benchmarks that we would like an operational Cuscore to approach as

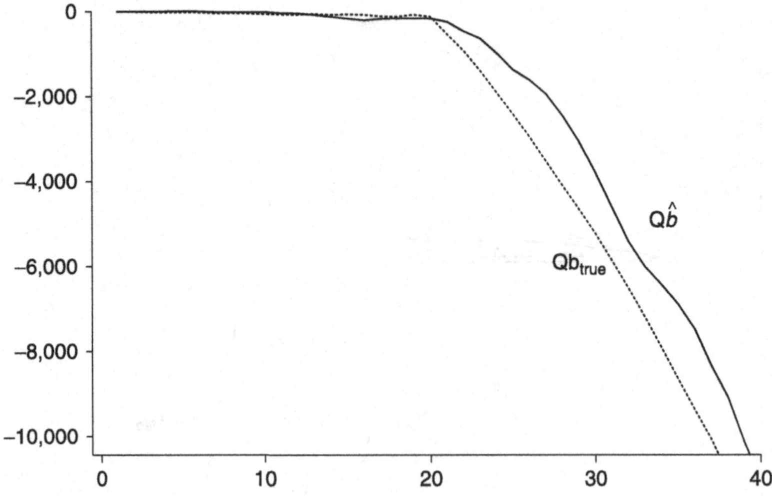

FIGURE 11.16 Cuscores for noisy data

closely as possible. Remarkably $Q\hat{b}$, our "product of two tardies" Cuscore, achieves an impressive standard of closeness with noiseless data. How is this possible? It is because larger discrepancies between the observation and the estimated local mean (because of the laggardly performance of the EWMA in the presence of sustained trending) are multiplied by a larger estimated slope following the change. The efficacy of the Cuscore is not so much the result of two tardies but of two enhanced discrepancies focused on a specific type of change. Does the scheme really work with price data? Look at Figure 11.16 and decide. And then think about how one might detect decreases in trend.

The presentation in this appendix is rather informal. For a rigorous treatment of dynamic modeling and identification of change, see Pole et al, 1994. In that reference, the linear growth model provides explicit parameterization of a local mean and trend (growth), dynamic updating of parameter estimates and forecasts, and formal statistical diagnostics for parametric (slope, in our case) change. The standard distributional assumptions of the DLM are strictly not appropriate for stock price data, largely because of notable nonnormality of (normalized) price and return distributions. Nonetheless, the models are useful if one takes a robust view of the formalities, concentrating on mean estimates, using standard deviations as a guide to uncertainty, and not counting on normality at all (so-called Linear Bayes methods).

Happy catastrophe hunting!

Bibliography

Arnold, V.I. *Catastrophe Theory*. New York: Springer-Verlag, 1986.

Bollen, N.P.B., T. Smith, and R.E. Whaley (2004). "Modeling the bid/ask spread: measuring the inventory-holding premium," *Journal of Financial Economics*, 72, 97–141.

Bollerslev, T. (1986). "Generalized Autoregressive Conditional Heteroskedasticity," *Journal of Econometrics*, 31, 307–327.

Box, G.E.P., and G. Jenkins. *Time Series Analysis: Forecasting and Control*. San Francisco: Holden-Day, 1976.

Box, G.E.P., and A. Luceno. *Statistical Control by Monitoring and Feedback Adjustment*. New York: John Wiley & Sons, 1987.

Carey, T.W. Speed Thrills. *Barrons*, 2004.

Engle, R. (1982). "Autoregressive Conditional Heteroskedasticity with Estimates of the Variance of United Kingdom Inflation," *Econometrica*, 50, 987–1,008.

Fleming, I. *Goldfinger*. London: Jonathan Cape, 1959.

Gatev, E., W. Goetzmann, and K.G. Rouwenhorst. "Pairs Trading: Performance of a Relative Value Arbitrage Rule," *Working Paper 7032, NBER*, 1999.

Gould, S.J. *The Structure of Evolutionary Theory*. Cambridge: Harvard University Press, 2002.

Huff, D. *How to Lie With Statistics*. New York: W.W. Norton & Co., 1993.

Institutional Investor. "Wall Street South," *Institutional Investor*, March 2004.

Johnson, N.L., S. Kotz, and N. Balakrishnan. *Continuous Univariate Distributions*, Volumes I and II. New York: John Wiley & Sons, 1994.

Lehman Brothers. *Algorithmic Trading*. New York: Lehman Brothers, 2004.

Mandelbrot, B.B. *Fractals and Scaling in Finance: Discontinuity, Concentration, Risk*. New York: Springer-Verlag, 1997.

Mandelbrot B.B., and R.L. Hudson. *The (Mis)Behavior of Markets: A Fractal View of Risk, Ruin, and Reward*. New York: Basic Books, 2004.

Orwell, George. *1984*. New York: New American Library, 1950.

Perold, A.F. (1988). "The Implementation Shortfall, Paper vs. Reality," *Journal of Portfolio Management*, 14:3, 4–9.

Pole, A., M. West, and J. Harrison. *Applied Bayesian Forecasting and Time Series Analysis*. New York: Chapman and Hall, 1994.

Poston, T., and I. Stewart. *Catastrophe Theory and its Applications*. London: Pitman, 1978.

Schack, J. (2004). "Battle of the Black Boxes," *Institutional Investor*, June 2004.

Sobel, D. *Longitude*. New York: Penguin Books, 1996.

Bibliography

Index

Accuracy issues, structural models, 59–61
Adaptive model, 172
Adjusted prices, 13n1
Advanced Theory of Statistics, The (Kendall, Stuart, and Ord), 63
Algorithmic trading (Black Boxes), 1, 3, 183–190
  dynamic updating, 188
  market deflation and, 189–190
  modeling transaction volume and market impact, 185–188
Alliance Capital, 165
Altvest, 161
American Airlines (AMR)–Continental Airlines (CAL) spread, 2, 10–16, 37–39, 40–45
Antilochus, 163
Applied Multivariate Analysis (Press), 65
"Arbed away" claim, 159–160
ARCH (autoregressive conditional heteroscedastic) models, 75–76
ARFIMA (autoregressive fractionally integrated moving average), 49
ARIMA (autoregressive integrated moving average), 48–49
Arnold, V. I., 205–206n3
Asian bird flu, 175
Autocorrelation, 129–130
Automatic trading, *see* Algorithmic trading (Black Boxes)
Autoregression and cointegration, 47–49
Autoregressive conditional heteroscedastic (ARCH) models, 75–76
Autoregressive fractionally integrated moving average (ARFIMA), 49
Autoregressive integrated moving average (ARIMA), 48–49
Avarice, catastrophe process and, 205–209

Ball bearing analogy, 211–212
Bamberger, Gerry, 1n1
Bank of America, 165, 185, 189
Barra model, 21
Barron's, 161
Bid-ask spread, declining, 156–159
Binomial distribution, 88–89
Black Boxes, 1, 3, 183–190
  dynamic updating, 188
  market deflation and, 189–190
  modeling transaction volume and market impact, 185–188
Block trading activity, 173
Bollinger bands, 17, 26
Bond futures, 85–87
Box, G.E.P., 1, 9, 48, 191
Brahe, Tyco, 6n3
British Petroleum (BP)–Royal Dutch Shell (RD) spread, 46–47

Calibration, 12–16, 13n1, 32–36
Carroll, Lewis, 67
Catastrophe process, 191–221, 205n3
  contrasted to popcorn process, 194–198
  Cuscore statistics and, 200–205, 211–221
  cusp, 206, 208
  forecasts with, 198–200
  move, 194–200
  normal factor, 206, 207
  risk management and, 209–211
  splitting factor, 206, 207
  surface, 205–206, 207
  theoretical interpretation of, 205–209
  trend change identification and, 200–205
Catastrophe Theory (Arnold), 205–206n3
Cauchy distribution, 74, 126
Change point identification, 200
Chi-square distribution, 96
Classical time series models, 47–52
  autoregression and cointegration, 47–49
  dynamic linear model, 49–50
  fractal analysis, 52
  pattern finding techniques, 51–52
  volatility modeling, 50–51

Cointegration and autoregression, 47–49
Competition, return decline and, 160–162
Conditional distribution, 118–119, 121–122
Conditional probability, 69
Consumer surplus, 159, 162
Continental Airlines (CAL)–American Airlines (AMR) spread, 2, 10–16, 37–39, 40–45
Continuity, 114–117
Correlation:
 first-order serial, 77–82
 during loss episodes, 151–154
Correlation filters, 21–22
Correlation searches, with software, 26
Covariance, 103
Credit crisis of 1998, risk and, 145–148
Credit Suisse First Boston, 26, 38–39, 185, 189
Cuscore statistics, 200–205, 211–221

Daimler Chrysler, 24
Debt rating, risk and, 145–148
Decimalization, 156–159
Defactored returns, 55–57, 65–66
D.E. Shaw, 3, 189
Difference, 48
Discount factor, 40, 43
Discrete distribution, 73, 115
Distributions:
 binomial, 88–89
 Cauchy, 74, 126
 Chi-square, 96
 conditional, 118–119, 121–122
 discrete, 73, 115
 empirical, 18
 Gamma, 96
 inverse Gamma, 76–77
 joint, 70–71
 lognormal, 134–135
 marginal, 86, 93
 nonconstant, 82–84
 normal factor, 17–18, 76–77, 82–84, 92–98, 120–124, 134
 sample, 123
 Student *t*, 75, 124–126, 201
 truncated, 123
 uniform, 82–84
Dividend, 13n1
DLM (dynamic linear model), 45, 49–50, 57
Dobzhansky, T., 155
Double Alpha, 3
Doubling, 11–12, 81–83
Dynamic linear model (DLM), 45, 49–50, 57
Dynamics, calibration and, 32–36
Dynamic updating, Black Boxes and, 188

Earnings, Regulation FD and, 150–151
EEG analogy, 200n1
Elsevier (ENL)–RUK spread, 99–101, 105
Empirical distribution, 18
Engle, R., 51, 75
Equilibrium, 192
ETFs (exchange traded funds), 181
European markets, decimalization and, 157–158
Event analysis, 22–26
Event correlations, 31–32
Event risk, 142–145
Evolutionary operation (EVOP), 32–36
EWMA (Exponentially weighted moving average), 40–47
 catastrophe process and, 202–204, 209–210
 Cuscore statistics and, 216–221
Exchange traded funds (ETFs), 181
Expected revealed reversion, 121–122
 examples, 123–124
Exponentially weighted moving average (EWMA), 40–47
 catastrophe process and, 202–204, 209–210
 Cuscore statistics and, 216–221
Extreme value, spread margins and, 16–18
ExxonMobil (XON)–Microsoft (MSFT) spread, 101, 104

Factor analysis, 54–55, 63–66
Factor model, 53–58
 credit crisis and, 147
 defactored returns, 55–57
 factor analysis, 54–55, 63–66
 prediction model, 57–58
Fair Disclosure Regulation, 150–151
Fear, catastrophe process and, 205–209
Federal Home Loan Mortgage Corp. (FRE)–Sunamerica, Inc. (SAI) spread, 142–145

Federal Reserve, 165
Feedback mechanism, 171–172
Fidelity, 163
Financial Times, 165
Flamsteed, John, 6n3
Ford (F)–General Motors (GM) spread, 101–102, 106–107
Forecast monitor, 42–43, 172
Forecasts:
  with catastrophe process, 198–200
  signal noise and, 167–174
Forecast variance, 26–28
Fractals, 52, 59, 73

Gamma distribution, 96
GARCH (generalized autoregressive conditional heteroscedastic), 50–51, 75–76
Gauss, 91
Generalized autoregressive conditional heteroscedastic (GARCH), 50–51, 75–76
General Motors, turning point example, 22–25
General Motors (GM)–Ford (F) spread, 101–102, 106–107
Geology analogy, 200n1
GlaxoSmithKline (GSK)–Pfizer (PFE) spread, 173
Goldman Sachs, 26, 185, 189
Gould, S. J., 174n1

Heavy tails, 17–18, 91–98
Hurdle rate of return, 209

IBM, 30
Inhomogeneous variances, 74–77, 136–137
Institutional Investor, 189
Institutional investors, return decline and, 163
Integrated autoregression, 48
Interest rates, 91, 211
  credit crisis and, 145–148
  volatility and, 165
International economic developments, risk and, 145–148
Interstock volatility, 67, 99–112, 164–165
Intervention discount, 43–45
Inverse Gamma distribution, 76–77
Iraq, U.S. invasion of, 175–176, 179

Janus, 46, 165
J curve, 62–63
Jenkins, G., 48
Johnson, N. L., 134
Joint distribution, 70–71

Kendall, Maurice, 63
Keynes, J. M., 91
Kidder Peabody, 150
Kotz, S., 134

Law of reversion, *see* Reversion, law of
Lehman Brothers, 26, 185, 189
Leptokurtosis, 73
Linear Bayes, 221
Liquidity, decimalization and, 158–159
Logistic function, 188
Lognormal distribution, 134–135
Long Term Capital Management (LTCM), 145, 150
Loss episodes, correlation during, 151–154
Luceno, A., 191

Managers:
  performance and correlation, 151–154
  relative inactivity of, 166–174
Mandelbrot, Benoit B., 52, 59
Marginal distribution, 86, 93
Market deflation, 189–190
Market exposure, 29–30
Market impact, 30–31, 185–188
Market neutrality, 29
Markowitz, Harry, 99
Marsh & McLennan, 180
Merck, 180
Mergers, return decline and, 161
Microsoft (MSFT)–ExxonMobil (XOM) spread, 101, 104
Miletus, 189
Mill's ratio, 123
Moment, sample, 17–18, 188
Momentum, 187–188
Morgan Stanley, 1, 3, 183–185
Morningstar, 165
Moving average, *see* Exponentially weighted moving average (EWMA)

Multiple bets, 11–12
Mutual funds, 176

National Bureau of Economic Research (NBER), 3, 18
NBER (National Bureau of Economic Research), 3, 18
Neural networks, 51–52
New risk factors, 145–148
New York Stock Exchange (NYSE), 3, 189
Noise models, 10–18
  multiple bets, 11–12
  reverse bets, 11
  rule calibration, 12–16
  spread margins, 16–18
Nonconstant distributions, 82–84
Nonfactor models, credit crisis and, 148
Nonmedian quantile movements, 135–136
Nonstationary processes, 136–137
Normal factor distribution, 17–18, 76–77, 82–84, 92–98, 120–124, 134
NYSE (New York Stock Exchange), 3, 189

Observational rules, xvn1, 10–18, 37–39
  calibration, 12–16
  spread margins, 16–18
Ord, Keith, 63
Outliers, 106, 117, 129

Pair identification, 20–26
Pairs trading, 1–3, 9–10
Partial autocorrelation, 129
Pattern finding techniques, 51–52. *See also* Algorithmic trading (Black Boxes)
PCA (principal component analysis), 54
Pfizer (PFE)–GlaxoSmithKline (GSK) spread, 173
Poisson process, 51
Popcorn process, 18–20, 58, 92
  contrasted to catastrophe process, 194–198, 205n3, 209–210
Prediction model, 57–58
Press, S. J., 65
Principal component analysis (PCA), 54
Probability, conditional, 69
Process adjustment, 211
Pure reversion, 118–120
  examples, 122–123, 124, 126–135, 139–140

Redemption tension, 148–150
Regulation Fair Disclosure (FD), 150–151
Relative price movement, *see* Interstock volatility
Resilience, 147n1
Return decline, 155–181. *See also* Return revival; Risk scenarios
  "arbed away" claim, 159–160
  competition and, 160–162
  decimalization and, 156–159
  institutional investors and, 163
  structural change and, 179–180
  temporal considerations, 166–174
  2003 and, 178–179
  volatility and, 163–165
  world events and business practices and recent, 174–178
Return revival, 191–221. *See also* Return decline
  catastrophe process, 194–198
  catastrophe process forecasts, 198–200
  catastrophe process theoretical interpretation, 205–209
  Cuscore statistics and, 200–205, 211–221
  risk management and, 209–211
  trend change identification, 200–205
Revealed reversion, *see* Expected revealed reversion
Reverse bets, 11
Reversion, law of, 67–89, 113–114, 139–140
  first-order serial correlation and, 77–82
  inhomogeneous variances and, 74–77
  interstock volatility and, 67, 99–112, 164–165
  looking several days ahead and, 87–89
  nonconstant distributions and, 82–84
  in nonstationary process, 136–137
  serial correlation, 138–139
  75 percent rule and, 68–74
  in stationary random process, 114–136
  temporal dynamics and, 91–98
  U.S. bond futures and, 85–87
Reynders Gray, 26
Risk arbitrage, competition and, 160–161
Risk control, 26–32
  event correlations, 31–32
  forecast variance, 26–28
  market exposure, 29–30
  market impact, 30–31

Risks scenarios, 141–154
catastrophe process and, 209–211
correlation during loss episodes, 151–154
event risk, 142–145
new risk factors, 145–148
redemption tension, 148–150
Regulation Fair Disclosure (FD), 150–151
Royal Dutch Shell (RD)–British Petroleum (BP) spread, 46–47

S&P (Standard & Poor's):
S&P 500, 28
futures and exposure, 21
Sample distribution, 123
Santayana, George, 5n2
SARS (severe acute respiratory syndrome), 175
Securities and Exchange Commission (SEC), 3, 150–151
Seismology analogy, 200n1
September 11 terrorist attacks, 175
Sequentially structured variances, 136–137
Sequentially unstructured variances, 137
Serial correlation, 138–139
75 percent rule, 68–74, 117
first-order serial correlation and, 77–82
inhomogeneous variances and, 74–77, 136–137
looking several days ahead and, 87–89
nonconstant distributions and, 82–84
U.S. bond futures and, 85–87
Severe acute respiratory syndrome (SARS), 175
Shackleton, E. H., 113
Sharpe ratio, 116
Shaw (D.E.), 3, 189
Shell, *see* Royal Dutch Shell (RD)–British Petroleum (BP) spread
Sinusoid, 19–20, 170
Spatial model analogy, 200n1
Specialists, 3, 156–157
Speer, Leeds & Kellog, 189
Spitzer, Elliot, 176, 180
Spread margins, 16–18. *See also specific companies*
Standard & Poor's (S&P):
S&P 500, 28
futures and exposure, 21
Standard deviations, 16–18
Stationarity, 49, 84–85
Stationary random process, reversion in, 114–136
amount of reversion, 118–135
frequency of moves, 117
movements from other than median, 135–136
Statistical arbitrage, 1–7, 9–10
Stochastic resonance, 20, 50, 58–59, 169, 204
Stochastic volatility, 50–51
Stock split, 13n1
Stop loss, 39
Structural change, return decline and, 179–180
Structural models, 37–66
accuracy issues, 59–61
classical time series models, 47–52
doubling and, 81–83
exponentially weighted moving average, 40–47
factor model, 53–58, 63–66
stochastic resonance, 58–59
Stuart, Alan, 63
Student *t* distribution, 75, 124–126, 201
Sunamerica, Inc. (SAI)–Federal Home Loan Mortgage Corp. (FRE) spread, 142–145

Tail area, 17–18
Takeover announcements, event risk and, 142–145
Tartaglia, Nunzio, 1–2, 11–12
Temporal considerations:
Black Boxes and, 185–188
return decline and, 166–174
Time weighted average price (TWAP), 193
Transaction volume, Black Boxes and, 185–188
Truncated distribution, 123
Turning point algorithm, 22–25
Turtle trade rule, 11, 20
TWAP (time weighted average price), 193

Uniform distributions, 82–84
U.S. bond futures, 85–87
Utility function, 38

Variances, inhomogeneous, 74–77, 136–137
Vector, 189

Vioxx, 180
Virgil, 183
Volatility:
  Black Boxes and, 189–190, 193–194
  catastrophe process and, 209–211
  interstock, 67, 99–112, 164
  measuring spread, 108–112
  return decline and, 163–165
Volatility bursts, 75–76
Volatility modeling, 50–51
Volume patterns, 24–25
VWAP (volume weighted average price), 163, 165, 193

Wavelet analysis, 51–52
Wilde, Oscar, 37
Wilmott, Paul, 1n1
World events, return decline and, 174–178